ENVIRONMENT & POLICY

VOLUME 57

More information about this series at http://www.springer.com/series/5921

Phoebe Koundouri
Editor

The Ocean of Tomorrow

The Transition to Sustainability – Volume 2

Springer

Editor
Phoebe Koundouri
School of Economics and ReSEES
Research Laboratory
Athens University of Economics
and Business; UN SDSN Europe;
EIT Climate KIC
Athens, Greece

ISSN 1383-5130 ISSN 2215-0110 (electronic)
Environment & Policy
ISBN 978-3-030-56845-0 ISBN 978-3-030-56847-4 (eBook)
https://doi.org/10.1007/978-3-030-56847-4

This Springer imprint is published by the registered company Springer Nature Switzerland AG
The registered company address is: Gewerbestrasse 11, 6330 Cham, Switzerland

During the last 25 years, I have written and published many scientific books and papers on holistic and interdisciplinary approaches that can enable a science-driven transition to a sustainable interaction between people, the economy and nature, while under the existential threat of Climate Change. I have invested considerable time and effort in mobilizing the transition to sustainability through science and innovation, and the engagement of businesses, policy making politicians and the civil society.

My inspiration and drive derive from my children Chrysilia, Billie and Athena, my strength and resilience from Nikitas. All books I have written in the past and will write in the future, are and will be dedicated to them.

Billie is the artist of the paintings displayed on the dedication and title pages of this book. I thank her for allowing us to see the "Oceans of Tomorrow" through her 8-year-old imagination!

Phoebe Koundouri
May 2020, Ecali, Athens

Oceans of Tomorrow
OCEANS
Oceans

Acknowledgements

The Assistant Editor of this book is Lydia Papadaki. Without her excellent editorial work and dedication to this book project, this book would not have been completed. I am in debt to her.

My gratitude goes to all the world-class scientists, from all across Europe and beyond, who contributed to the 13 interdisciplinary research and demonstration projects as well as the three global initiatives that form the basis of this book[1]: two FP7 European Commission DG Research and Innovation projects, MERMAID and TROPOS; two H2020 European Commission DG Research and Innovation projects, BLUEBRIDGE and COASTAL; three European Commission Interreg projects, AMARe, RECONNECT and Thal-Chor; six European Institute of Innovation and Technology[2] Climate KIC[3] projects, BL.EU. Climate, Deep Demonstration Project in European Ports, MEDFreeSup, Circular Learning Hub, Circular Economy (CE) in Smart Specialization Strategy (S3), and Western Balkan Circular and Climate Innovation-Beacons; as well as the three Global Initiatives, the 4-Seas UN SDSN[4] Euro-Asian Initiative, the UN SDSN Global Roundtable for Sustainable Shipping and Ports, and the United Nations Initiative for Climate Change Effects on Cultural and Natural Heritage.

These 16 projects and initiatives, some completed but the vast majority ongoing, form the "Sustainable Blue Growth" research domain of the cluster of the institutions I direct, which includes the ReSEES Research Laboratory at the Athens University of Economics and Business,[5] EIT Climate KIC Hub Greece[6] at the ATHENA

---

[1]Here I refer to the acronyms and the funding instrument of each project. Details for each project can be found in the main text of the book.

[2]EIT, European Institute of Innovation and Technology.

[3]https://www.climate-kic.org/

[4]http://www.unsdsn.gr

[5]https://www.dept.aueb.gr/en/ReSEES

[6]https://www.athenarc.gr/en/eit-climate-kic-greece-hub

Research and Innovation Center, and the United Nations Sustainable Development Solutions Network Greece. As a response to the climate crisis and its effect on marine ecosystems and coastal populations, the Sustainable Blue Growth domain is one of the many research domains of our cluster and aims at establishing transformation pathways towards Sustainable Blue Growth, which will be supported by technically and socially innovative solutions. We focus on research, education, development and support of open access databases, innovation incubation and acceleration, and policy recommendations, while our aim is to engage all relevant stakeholders in co-designing a systems innovation pathway for the transition to socially, culturally, economically, environmentally and geopolitically sustainable development in Europe and beyond, embracing seas and oceans.

This book has been finalized during the COVID-19 pandemic. This pandemic has clearly proven the ability of governments to take dramatic measures to mitigate an existential threat, as well as people's ability, at least in the short run, to adapt to new restricted lifestyles imposed by these measures. A second message is that the timing of the enactment of measures is crucial for their effectiveness in saving lives. A third message is that the response to COVID-19 came from national states, while international organizations lack in terms of explicit imminent response. Importantly, there is serious scientific speculation that COVID-19 might be connected to the climate crisis and the related loss in biodiversity. Deforestation drives wild animals closer to human populations, increasing the likelihood that zoonotic viruses will make the cross-species leap. Moreover, the Intergovernmental Panel on Climate Change warned that global warming will likely accelerate the emergence of new viruses.

What one cannot help but notice is that the response to the COVID-19 pandemic is very different to the lack of effective action on climate change, the other existential crisis of our times. One should ask why. Climate change has the potential to end up killing more people than COVID-19, but the deaths reference of this crisis is hidden in the jargon as "increased frequency and severity of natural disasters" and is spread over decades. IPCC 2018 reports that the level and speed of the change needed, to successfully tackle the climate crisis, is unprecedented. Incremental changes will not be enough!

Our generation has lived to see at least three global crises: the financial crisis of 2007–08, the COVID-19 pandemic and the unprecedented economic crisis (the pandemic has created a demand shock, a supply shock, and a financial shock all at once) deriving from the required social-distancing measures taken to contain the spread the virus, and the mother of all crises, the climate crisis. If we continue attempting to face each new crisis with the same socio-economic model that gave rise to the crisis, we will fail to find a sustainable and resilient socio-economic–environmental pathway. In downturns, as Darwin surmised, those who survive *"are not the strongest or the most intelligent, but the most adaptable to change"*.

I believe that we can do even better than just react to crises by adapting to the new crisis-born reality. We can use science – as we are using science currently for designing measures to restrain the diffusion of COVID-19 – to design economies that will mitigate the threats of climate change, biodiversity loss and pandemics. We need to leverage the power of people to achieve the vision of a prosperous, inclusive, climate and pandemic resilient society with a circular, net-zero emissions economy. The IPCC report explicitly refers to the need for "rapid far-reaching and unprecedented changes in all aspects of society". Incremental changes will not be enough. What is needed now is a fundamental transformation of economic, social and financial systems that will trigger exponential change in strengthening social, economic, health and environmental resilience. We need big thinking and big changes. System innovation and transitions thinking can help and calls for intense public participation.

Following the 2008 financial crash, we saw public funds flow disproportionately to polluting industries and to society's most wealthy. This must not happen again. We must start investing in what makes our socio-economic system resilient to crisis, by laying the foundation for a green, circular economy that is anchored in nature-based solutions and geared towards public well-being. Now is the time to usher in systemic economic change and the good news is that we have our blueprint: it's the combination of UN Agenda 2030 (17 SDG) and European Commission's European Green Deal. Now is the time for financial institutions and governments to embrace EU taxonomy for sustainable investments (2019), to phase out fossil fuels by deploying existing renewable energy technologies, eliminate fossil fuel subsidies – amounting to 5.2 trillion per annum – and redirect them to green and smart climate mitigation and adaptation infrastructural projects, invest in circular and low carbon economies, shift from industrial to regenerative agriculture, exploit the limits of the digital revolution, and reduce transportation needs.

A decisive march along this sustainable pathway will enhance economic and environmental resilience, create jobs, and improve health and well-being in both rural and urban communities. The transition should be inclusive and "leave no one behind", that is why finance should be directed not only to those who are sustainable or have the potential to become sustainable, but also to those who are willing to commit and be monitored henceforth, to learning how to become sustainable.

Such a decisive march is of upmost importance for our seas and oceans!

Never waste a good crisis!

Ecali, Athens
May 2020

Prof. Dr. Phoebe Koundouri

Contents

About the Editor and Contributors

The Editor

Phoebe Koundouri holds a PhD and MPhil in Economics and Econometrics from the University of Cambridge (UK). She is Professor (Chair) of Economics in the School of Economics, Athens University of Economics and Business (Greece), and she is the elected President of the European Association of Environmental and Natural Resource Economists (EAERE) (with more than 1200 scientific member institutions, from more than 75 different countries). Prof. Koundouri is listed in the most-cited women economists in the world, with 15 published books and more than 300 peer reviewed scientific papers.

Prof. Phoebe Koundouri is the Founder and Scientific Director of the Research laboratory on Socio-Economic and Environmental Sustainability (ReSEES, https://www.dept.aueb.gr/en/ReSEES) at Athens University of Economics and Business. She is also Affiliated Professor at the ATHENA Research and Innovation Center (Greece), where she directs EIT Climate-KIC Hub Greece (https://www.climate-kic.org/) of the European Institute of Innovation and Technology, focusing on accelerating innovations for the transition to a climate neutral economy. She is also the Co-chair of the United Nations Sustainable Development Network Europe (UN SDSN Europe, https://www.unsdsn.org/), Chair of the Scientific Advisory Board of the International Centre for Research on the Environment and the Economy (ICRE8, http://icre8.eu/), and Chair of the Scientific Advisory Board of the European Forest Institute (https://efi.int/).

In the past, Professor Koundouri has held academic positions at the University of Cambridge, University College London, University of Reading and London School of Economics. She acts as an advisor to the European Commission, World Bank, EIB, EBRD, OECD, UN, NATO, WHO, etc. numerous national and international foundations and organizations, as well as national governments in all five continents. She is one of the Commissioner of the Lancet Commission on COVID-19 (https://covid19commission.org/commissioners) for which she co-chairs the "Jobs-based Green Recovery Task Force", she co-leads the UN SDSN Sustainable Ports and

Shipping Initiative, the UN SDSN 4-seas (Mediterranean, Black, Caspian and Aral Seas) Blue Growth Initiative and the UN SDSN Senior Working Group on "Transformation Pathways for the implementation of EGD and the SDGs". She is a member of the CEPR (Center for Economic Policy Research) Network (RPN) on Climate Change https://cepr.org/content/cepr-rpn-climate-change-researchers, member of the Priministerial Committee for the Recovery and 10-year Development Plan of Greece, the Climate Change Committee of the Greek Ministry of Environment and Energy, as well as numerous European and International Scientific, Research and Policy Boards and Committees. Since 1997, she has coordinated more than 100 interdisciplinary research projects and has attracted significant competitive research funding. Professor Koundouri and her large interdisciplinary team have produced research and policy results that have contributed to shaping European policies. Over the last two decades, Professor Koundouri has given keynote and public lecturers all over the world and received various prizes for academic excellence, including the prestigious European Research Council (ERC) Synergy Grant (2021–2027).

Contributors

Ebun Akinsete is a Senior Researcher at ICRE8 and the Head of the Department of Stakeholder Analysis and Decision Support Systems at ICRE8 and UN SDSN Greece. She is an Associate Lecturer in the School of Applied Social Studies at Robert Gordon University (Aberdeen) and Senior Partner with Nigeria-based consultancy GEN Sustainable Solutions. With a BSc (Hons) in Architectural Technology, MSc in Project Management and PhD in Urban Regeneration and Sustainable Communities, she has several years of both industrial and academic experience in Europe, Asia and Africa, collaborating with governmental agencies, NGOs, aid organizations, academic institutions, the private sector and local communities. Her main areas of interest are sustainable development, urban regeneration, SDG policy and implementation, community development, climate change mitigation, transition management, renewable energy, stakeholder participation, participatory planning, evaluation and impact assessment for sustainability, and sustainability in developing nations.

Vera Alexandropoulou is a graduate from the Law Faculty at the University of Athens and has completed postgraduate studies at the University of Heidelberg and Harvard University. She was admitted to the Greek (Piraeus) Bar Association in 1998, qualified as UK Solicitor in 2004 and obtained her Cyprus Bar registration in early 2017. Vera has worked for the Harvard Business School, FHM – a major shipping law firm in New York, and Norton Rose and is General Counsel at the Piraeus Port Authority. She established the Alexandropoulou Law Firm in 2005. Vera has acquired significant experience in sectors including commercial and corporate law, energy, banking and finance, and real estate. She is involved as Chief

Legal Officer with various institutions and associations, mainly in the field of energy business and fund structures, and is the author of codification of Greek renewable energy and natural gas legislation, having published articles on Greek ship finance, Greek listed funds and environmental issues. Vera has also been certified as DPO Executive by TUV Austria Hellas. She is fluent in Greek, English, French, German and Spanish and understands Chinese.

Gerasimos Antzoulatos received his BSc Degree in Mathematics in 1999 and MSc Degree in "Computer Mathematics and Decision Making" from the University of Patras in 2002. Since January 2018, he works as a Research Associate at the Information and Technologies Institute, Centre for Research and Technology Hellas (CERTH/ITI). His research interests include Computational Intelligence methods and their application to Machine Learning, Business Intelligence and Predictive Analytics. He worked as Lecturer in many Departments at Higher Technological Educational Institute of Patras and as a Consultant at Data Research & Consulting Company S.A. From 2015 until 2018, he worked in the Research and Development Department at I2S SA where his main duties included design, development and evaluation of data analytics technologies and tools in aquaculture and maintenance field. Also, he was a member of a research team working on two prestigious Horizon2020 projects, BlueBRIDGE and AquaSmart.

Maria D. Argyrou holds a BA in Economic Sciences from the University of Athens, Greece (graduated Cum Laude and was first in class), and a Master of Public Administration (Development Practice and specialization in Advanced Policy and Economic Analysis) from Columbia University, New York City, USA, where she has been awarded two scholarships (Paul Nichoplas Scholarship and Gerondelis Foundation Scholarship). She has worked as a research assistant in the Department of Economics at the University of Athens as well as in the School of International and Public Affairs at Columbia University. She has also worked as a Researcher at the International Center for Biosaline Agriculture (ICBA) in Dubai and for the European Bank for Reconstruction and Development as SIPA Capstone Workshop Consultant.

Wenting Chen holds a PhD degree in Natural Resource Management and a Master's degree in Environment and Development Economics. Dr. Chen was Guest Researcher at the University of California, Berkeley. She has worked extensively on subjects related to marine ecosystem service and sustainable ocean management with focus on valuation and modelling of marine ecosystem services, sustainable marine resource management, sustainable ocean development and governance, blue growth, integrated marine habitat, and coastal management. Wenting also has good experience on environmental policy analysis, institutional economics, payment for environmental services, governance of public goods and the commons. She has a leading role on socioeconomic impact analysis in several EU, international and national projects including AMAP, Horizon 2020 MERCES, EU FP7 TROPOS,

EU FP7 MERMAID, NFR JELLEYFARM and MIKON ECOURCHIN. She is also involved in other EU projects such as Horizon 2020 TAPAS and ResponSEAble og EU FP7 MARS.

Chia-Fa Chi is a Doctoral Candidate in the Department of Marine Environment and Engineering at the National Sun Yat-sen University, Taiwan. His main research activities are focused on adaptation to climate change in coastal areas. In particular, he is interested in the issues of maladaptive risks. He has received awards of the Graduate Students Study Abroad Program, which is sponsored separately by Taiwan Ministry of Science and Technology and the National Sun Yat-sen University.

Osiel González Dávila is Research Associate at the Athens University of Economics and Business and Athena–RC. He has worked as Guest Teacher at the London School of Economics and Political Science and as Lecturer in Economics at SOAS, University of London. He holds an MRes in Environment and Development from the University of Lancaster (2008). His principal research interests lie in the fields of economics, environment and development and he has collaborated on the following EU-funded research projects: MERMAID, THESEUS, TROPOS and OpenAIRE.

Fernando Del-Jesus is Civil Engineer specialized in Offshore Wind Energy. His PhD Thesis entitled "Analysis of climate variability influence on offshore wind farms" contributes to offshore wind characterization, floating platform design and economic feasibility of offshore wind farms. He has published in the main journals of the sector, such as *Wind Energy*, and he has participated in several conferences, OMAE, EWEA, and Offshore, among others. Fernando has also been part of INORE (International Network for Offshore Renewable Energy) as Committee Member for 3 years acting as Research Manager.

Elias Demian is Research Associate at the Foundation for Economic and Industrial Research. He holds a BSc degree in Environmental Science and an MSc degree in Environmental Management and Policy from Lund University, International Institute of Industrial Environmental Economics, Sweden. Employed in completion and management of European programmes implemented nationally, in the Greek shipping sector (Department of Environmental Management and Law), he is at present a freelancer Environmental Consultant. His interests include product policy issues, cleaner production methods, industrial ecology, inter-organizational management schemes and environmental management systems (EMAS, ISO 14001).

Charalampos Dimitrakopoulos holds an MSc in Banking and Finance from the University of Piraeus and a Bachelor's degree in Mathematics from the University of Patras. Currently, he is an IT Consultant at Communication & Information Technologies Experts SA and has worked on Market and Operational Risk at Alpha Bank.

Raúl Guanche, PhD, MSc Civil Engineering, is responsible for the Offshore Engineering and Marine Renewable Energies Research Group at the Environmental Hydraulics Institute, University of Cantabria. He has had an intense research career and collaborated with engineering firms through the development techno-scientific projects in different areas related with the ocean engineering sector. In the last 8 years he has participated in more than 50 projects with national and international companies in the field of offshore engineering and more specifically in the marine renewable energy sector, having developed tools and methodologies for a large number of companies. He has more than 50 techno-scientific publications, most of them in specialized journals like *Ocean Engineering*, *Renewable Energy* and *Wind Energy*. He has co-authored eight patents on devices linked to the marine renewable energy and offshore aquaculture sector.

Lars Golmen is a Senior Research Scientist at the Norwegian Institute for Water Research. He has expertise in fjord and coastal oceanography, including measurements, modelling and environmental assessments. As a partner in the Runde Environmental Centre and Manager of the Wave-Energy Test Site, he is part of very exciting times in terms of wave energy development.

Alice Guittard is a Geographer specialized in Sea & Coastal Management and GIS & Remote Sensing, currently holding a position as a Junior Researcher at ICRE8. She holds a Master's Degree in Geography specialized in Sea & Coastal Management from the University of Paul Valéry (Montpellier – France) and a Master's Degree in GIS & Remote Sensing from Stockholm University (Sweden). Her main research and interest focus on land–sea interactions, marine spatial planning, integrated coastal zone management, wetland management and ecosystem services. She is currently involved in the EU H2020 COASTAL project. She also worked for the CNRS (French National Research Agency) on EU-funded projects related to port sustainable development and transports. Alice has participated as a coach on systems innovation at the Pioneers into Practice workshops held in Greece.

Claire Haggett is Lecturer in Sociology of Sustainability. Her research focuses on the policy, planning and politics of renewable energy development, particularly the role of public perceptions, community engagement and landscape impact. She has been Project Manager for a series of research projects for the Scottish Government and ClimateXChange, the latest of which have explored community investment in commercial energy schemes and community benefits for offshore renewables. She is also currently working with Local Energy Scotland on the Good Practice Principles for Community Engagement for onshore and offshore wind. Claire is currently the Co-investigator on an EU Framework 7 project on offshore energy, leading on social acceptance issues. She works closely with the UK and Scottish Governments, Marine Scotland, the Crown Estate, and key stakeholders and policy makers from the Northern Ireland and Danish Governments on social responses to renewables and is the invited Expert for the International Energy Agency's work on the social acceptance of wind energy.

Klimanthia Kontaxaki is a registered Lawyer at the Athens Bar Association and graduate in the Faculty of Law at the National Kapodistrian University of Athens. Having a particular interest in the fields of Law of the Sea and International Shipping Law, in the summer of 2014, she was awarded a full scholarship by the Aegean Institute of the Law of the Sea and Maritime Law for attending the summer session courses (2014) in "Ocean Management and Maritime Law" at Tulane University/New Orleans (USA) summer school in Rhodes. She is a holder of an LLM degree in Public International Law with dissertation in "Maritime Spatial Planning" (2016) and an LLM degree in International Shipping Law from the Queen Mary University of London (2018) under a scholarship awarded by the Centre of Commercial Law Studies (CCLS) at Queen Mary University of London.

Shiau-Yun Lu is an Associate Professor of Marine Environment and Engineering at the National Sun Yat-sen University in Taiwan. Her research focuses on the human settlement and influence in the interface between marine and territorial environment. The research specialization recently includes coastal resilience cities, marine and coastal management, and port city development. She received her Master's degree (in Landscape Architecture) from the University of Pennsylvania and her Doctoral degree (Doctor of Design) from Harvard University. She is LEED AP (Leadership in Energy and Environmental Design Accredited Professional) certificated since 2009. She was awarded the ASLA Professional Excellence in Analysis and Planning Category (2011) and the commendation prize in the Next Generation Container Port (NGCP) Challenge in Singapore (2013).

Evita Mailli is a Research Associate at MaDgik (Management of Data, Information, and Knowledge Group) in the Department of Informatics at the National and Kapodistrian University of Athens and at ICRE8 (International Centre for Research on the Environment and the Economy). She holds a BSc in Mathematics from the University of Patras and an MSc in Computer Science from the Faculty of Informatics and Telecommunications at the University of Athens. Her current interests involve data analysis, data visualization, and web-based decision support tools.

Vassiliki Manoussi holds a PhD degree in Economics with orientation of "Environmental Economics" from Athens University of Economics and Business (Department of International and European Economic Studies) and a Master's degree in Economics from Athens University of Economics and Business (Department of Economic Studies). She has worked as a Junior Researcher at Fondazione Eni Enrico Mattei (FEEM) and Centro Euro-Mediterraneo sui Cambiamenti Climatici (CMCC) and as a Postdoctoral Researcher at AUEB (Department of International and European Economic Studies). Her main research interests are Environmental Economics, Economics of Climate Change, Sustainable Development, Cost Benefit

Analysis, Economics of Risk and Uncertainty, Renewable Resources, Growth and Environmental Policy.

Saúl Torres-Ortega, PhD, MSc Civil Engineering, is a Lecturer in Business Administration since 2010 at the University of Cantabria. He is also a Researcher at the Environmental Hydraulics Institute. His research focuses on risk assessment, economic valuation of non-marketed resources and cost-benefit analysis (both environmental and public investments projects). He has participated in different national and international research projects and has published in several academic journals.

Lydia Papadaki is a PhD student at the Athens University of Economics and Business. She received her MSc in Economics and Policy of Energy and the Environment from the University College London (UCL) and her BSc (Distinction) in Economic Theory and Policy from Athens University of Economics and Business (AUEB). She is the Manager of UN SDSN Greece and EIT Climate-KIC Hub Greece, while she works as a Researcher in the following EIT Climate-KIC funded projects: Deep Demonstration of Maritime Systems Innovation in Piraeus, Climate Innovation in Southern European Waters, Circular Economy Transition in Smart Specialization Strategy in RIS Countries, and Circular Learning Hub. Previously, she worked as a Researcher at the International Centre for Research on the Environment and the Economy (ICRE8) at the DAFNE (H2020) project. Her main areas of interest are economics, environmental economics, renewable energy, climate change and sustainable development.

Andreas Papandreou is a Professor in the Department of Economics at the University of Athens. He received his BSc in Economics from the University of London, Queen Mary College, and his MSc in Economics from London School of Economics. He became a Doctor of Philosophy from Merton College, Oxford University, in 1990. Dr. Papandreou has 23 years of academic experience in various universities, including the University College London and Harvard. His research and published work are mostly in the fields of environmental economics, institutional economics and microeconomics. Among his publications are "*Externality and Institutions*" (1994).

Eleni Petra holds a Diploma in Computer Engineering and Informatics from Technical University of Patras, Greece, and a Master of Science in Computation from the University of Manchester, Institute of Science and Technology (U.M.I.S.T.) Her technical specialty is in Information Systems, User Interfaces, Human Factors in MMI, Data Base Management Systems, e-infrastructures and data infrastructures. Currently, she is working for Athena Research and Innovation Center and the University of Athens as Project Manager in Horizon2020 projects related to research infrastructures (RDA) and virtual research environments for marine environment and aquaculture (BlueBridge H2020 project).

David Rudolph is a Researcher at the Technical University of Denmark, Department of Wind Energy, Denmark. He has a PhD in Human Geography from the University of Edinburgh. His research focuses on Renewable Energy, Qualitative Research, Human Geography, Geography, Qualitative Methods, Discourse Analysis, Social Theory, Wind Energy, Communities, Public Engagement, Community Engagement, Marine Spatial Planning and Community Benefit. In his previous research, he has mainly employed a number of qualitative research methods, such as interviews, focus groups, content, policy and discourse analysis, to look at local responses, conflicts, policies, community engagement and benefits related to the planning and siting of onshore and offshore wind farms in the UK, Denmark, Ireland and Germany.

Pedro Díaz-Simal, PhD, MSc, Civil Engineering, MSc in Economics, is Professor of Economics in the School of Civil Engineering at the University of Cantabria. He specializes in environmental economic analysis and economic analysis of engineering. He is also a Senior Researcher at the Environmental Hydraulics Institute, a research centre developed at the University of Cantabria specialized in environmental issues. His main research areas are economics of climate change, economic analysis of ecosystem services and economic assessment of engineering projects impact.

Lydia Stergiopoulou is a Researcher in the Department of Sustainable Development at ICRE8 and Lead Scientist for the RECONNECT (Interreg Balkan-Med) project. She holds a Master's Degree in Environmental Sciences, Policy and Management from the University of Manchester and a Bachelor's Degree in Economics from the National and Kapodistrian University of Athens. Lydia works as a volunteer in field studies at the Ephorate of Underwater Antiquities, her previous position was at the Hellenic Centre for Marine Research on Blue Energy and has in-depth experience in Blue Growth and Blue Economy. Her research interests focus on natural and cultural ecosystem services valuation, blue tourism and sustainable development.

Marian Stuiver is a Senior Social Scientist with a focus on Sustainable Business Development, (Marine) Spatial Planning and Transition Management. She has experience in global (such as EU/UN) projects as well as local projects with multiple-stakeholders. Marian has extensive experience in guiding and understanding complex programme and project management of international and multidisciplinary teams. She has a rich toolbox of methodologies to connect representatives of industry, non-governmental organizations and policymakers from different backgrounds and cultures for a common goal. Her passion is to vision, develop and reflect on science for impact.

Stella Tsani holds a PhD in Economics and Business from the University of Reading, UK. She obtained her BSc degree in Economics from the National and

Kapodistiran University of Athens, Greece, and her MA degree in Business and Management in Emerging Markets from the University of Reading, UK. Her research interests focus on resource economics, energy, development, political economy, macroeconomics and socio-economic analysis. She has led research in projects funded by the European Commission (FP7, Horizon 2020) and various Directorates (DGENV, DGENER), the World Bank, Revenue Watch Institute USA, etc. Stella has worked for the Centre for Euro Asian Studies in the UK, Europrism in Cyprus, the Public Finance Monitoring Centre in Azerbaijan, the Bank of Greece, the Institute of Energy for South East Europe and the E3MLab at the National Technical University of Athens in Greece. She has worked for the UK Foreign Office Chevening Fellowship Program in Energy Economics hosted by the University of Reading. Stella held academic posts at Athens University of Economics and Business, the University of Reading, Kainar University and the Mediterranean College. She has published in leading peer-reviewed journals including: *Economics Letters*, *Resources Policy*, *Energy Economics* and *Economic Systems*.

Lena Tsipouri is Professor in the Department of Economic Sciences at the University of Athens and Senior Researcher in the Centre of Financial Studies of the Department. She studied Economic Sciences at the Universities of Athens and Vienna and completed her PhD (Doctorat d' Etat) at the University of Paris II, where received the first prize. She then undertook postdoctoral research at MIT under a Fulbright Fellowship. Prof. Tsipouri teaches Economic Development, European Economic, Integration, Economics of Technological Change and Theory of the Firm at both undergraduate and graduate level. Her scientific research as well as her presentations at various refereed scientific conferences and policy workshops are about Research and Innovation (focusing on start-up ecosystems), Regional Development and Corporate Governance.

Sander van den Burg is a Software Engineer and Researcher in software deployment, configuration management, software architectures, and free and open source software. He is a PhD student at the Software Engineering Research Group, Delft University of Technology, the Netherlands. His research was part of the Pull Deployment of Services (PDS) project and addresses software deployment issues in healthcare environments. He is working at Mendix, a company that provides a low-code application development platform.

Achilleas Vassilopoulos is an Assistant Professor in the Department of Economics at the University of Ioannina in Greece. He holds an MBA and a PhD in Consumer Theory from the Department of Agricultural Economics at the Agricultural University of Athens (AUA) and has served as a Visiting Scholar at the University of Arkansas, USA (Dale Bumpers College, Dept. of Agricultural Economics & Agribusiness). Achilleas has worked as a Consultant Economist for the Food and Agriculture Organization of the United Nations (UN FAO) and the Innovation and

Entrepreneurship Unit of AUA and has participated in more than ten EU-funded projects on behalf of both public and private institutes in Greece and abroad. His research interest focuses on the analysis of economic behaviour using quantitative and experimental methods with an emphasis on the agro-food sector and the environment. His academic work has been published in journals such as *Economic Modelling*, *Economics Letters*, *Journal of Applied Statistics*, *Journal of Consumer Affairs*, *European Review of Agricultural Economics* and others.

Petros Xepapadeas is a Postgraduate Researcher in Operations Research at the Athens University of Economics and Business. He received his BSc in Informatics from Athens University of Economics and Business and his MSc in Business Mathematics from a joint program at the National and Kapodistrian University of Athens and Athens University of Economics and Business. He has experience in risk analysis and has worked on the Open AIRE, MERMAID and BRIGAID projects.

Yung-hsian (Frank) Ying is Dean at the Office of International Affairs, National Taiwan Normal University. He holds a PhD in Economics from the University of Kentucky and his research interests are International Finance, International Economics, Political Economy and Macroeconomics.

Jason Yu is Assistant Professor at the National Sun Yat-sen University, Taiwan. His research interests are modelling of marine systems (tides, currents, storm surges, water qualities and ecology), water treatment, pipeline analysis and sewer design, and parallel computation. His recent research projects are study of the water quality improvements in fishery harbours, COA; study of the thermocline variabilities due to tide and seasonal variations in the Kaoping coastal seas, NSC; storm surge simulation on parallel computers, NSC; review of the hydrological instruments in Taiwan, WRB; oil spill modelling, CPC; and evaluation and construction oil spill models for the seas surrounding Taiwan, EPA.

Abbreviations

AAPA	American Association of Port Authorities
AIVP	International Association of Cities and Ports
ASPF	Arab Sea Ports Federation
CAP	Common Agricultural Policy
CAPEX	Capital Expenditure
CBA	Cost Benefit Analysis
CE	Choice Experiment
CES	Cultural Ecosystem Services
CGFI	China Green Freight Initiative
CICES	Common International Classification of Ecosystem Services
CO_2 eq	Carbon Dioxide Equivalent
CU	Central Unit
CV&C	Climatic Variability and Change
DECEX	Decommissioning Expenditure
DJSI	Dow Jones Sustainability Index
DSS	Decision Support System
EC	European Commission
ECSA	European Community Shipowner's Association
EDSNA	Association of Municipalities in the Attica Region – Solid Waste Management
EEA	European Environment Agency
EEDI	Energy Efficiency Design Index
EIA	Environmental Impact Assessment
EIB	European Investment Bank
EIT	European Institute of Innovation and Technology
EGD	European Green Deal
EMAS	Eco-Management and Audit Scheme
EMS	Environmental Management System
EOAN	Greek Recycling Organization
EPA	Environmental Protection Agency

ES	Ecosystem Services
ESD	Effort Sharing Decision
ESDAK	Association of Solid Waste Management of Crete
ESG	Environmental, Social and Governance
ESI	Environmental Ship Index
ESPO	European Sea Ports Organization
ETS	European Trading System
ERS	Emissions Reduction Strategy
ESIF	European Structural and Investment Funds
ETD	Energy Taxation Directive
EU	European Union
EYSSA	Special Service for Strategy, Planning and Evaluation
FCR	Feed Conversion Rate
FDR	Financial Discount Rate
FNPV	Financial Net Present Value
FP7	Seventh Research Framework Programme
FR	Feeding Rate
FRR	Financial Rate of Return
GAMs	Generalized Additive Models
GDP	Gross Domestic Product
GHG	Greenhouse Gas
GloMEEP	Global Maritime Energy Efficiency Partnerships
GPA	Global Programme of Action
H2020	Horizon 2020
IAPH	International Association of Ports and Harbours
ICCT	International Council for Clean Transportation
ICS	International Chamber of Shipping
IMarEST	Institute of Marine Engineering, Science and Technology
IMO	International Maritime Organization
IRR	Internal Rate of Return
IT	Information Technology
Kg	Kilogram
KIC	Knowledge and Innovation Community
KPIs	Key Performance Indicators
KWh	Kilo Watt Hour
LCoP	Levelized Cost of Production
LNG	Liquefied Natural Gas
LPG	Liquefied Petroleum Gas
MAC	Marginal Abatement Cost
MAES	Mapping and Assessment of Ecosystem Services
MALs	Multi-Actor Labs
MARS	Multivariate Adaptive Regression Splines
MEA	Millennium Ecosystem Assessment
MEPC	Maritime Environment Protection Committee

MPAs	Marine Protected Areas
MS	Member States
MSFD	Marine Strategy Framework Directive
MSP	Marine Spatial Planning
MUPs	Multi-use Platforms
MUOPs	Multi-use Offshore Platforms
MW	Mega Watt
NCES	National Circular Economy Strategy
NGOs	Non-governmental Organizations
NO_X	Nitrogen Oxide
NPV	Net Present Value
NPWM	National Plan for Waste Management
OECD	Organization for Economic Co-operation and Development
OPEX	Operating Expense
OP	Operational Programmes
OPS	Onshore Power Supply
OSWM	Organizations of Solid Waste Management
OTEC	Ocean Thermal Energy Conversion
PERS	Port Environmental Review System
PIANC	World Association for Waterborne Transport Infrastructure
PMAWCA	Port Management Association of West and Central Africa
PWP	Plastic Waste Prevention
RWMP	Regional Waste Management Plans
R&I	Research and Innovation
SCBA	Social Cost-Benefit Analysis
SD	Systems Dynamics
SDM	Self Diagnosis Method
SDGs	Sustainable Development Goals
SDSN	Sustainable Development Solutions Network
SEA	Significant Environmental Aspects
SED	Systems of Alternative Management
SNPV	Social Net Present Value
SEEMP	Ship Energy Efficiency Management Plan
SMEs	Small and Medium Enterprises
SO_X	Sulphur Oxide
SPAMI	Specially Protected Area of Mediterranean Importance
SSS	Smart Specialization Strategy
SUP	Single-Use-Plastic
SUPD	Strategy for Plastics in a Circular Economy
TEAL	Transport, Energy, Aquaculture and Leisure
TEU	Twenty-Foot Equivalent Unit
TEV	Total Economic Value
TIPC	Taiwan International Port Corporation
TRL	Technology Readiness Level

UK NEAFO	UK National Ecosystem Assessment and its follow-on phase
UN	United Nations
UNCTAD	United Nations Conference on Trade and Development
UNEP	United Nations Environment Programme
UNFCCC	United Nations Framework Convention on Climate Change
VAT	Value Added Tax
VREs	Virtual Research Environments
WFD	Water Framework Directive
WPs	Work Packages
WPSP	World Ports Sustainability Program
WTA	Willingness to Accept
WTP	Willingness to Pay

Chapter 1
Introduction to the Oceans of Tomorrow: The Transition to Sustainability

Phoebe Koundouri, Vassiliki Manoussi, and Lydia Papadaki

Abstract Science and technology offer an opportunity to reconcile the protection of marine ecosystems with the development of sustainable maritime activities, through an integrated maritime policy. In this context, the European Commission has developed a strategy with the aim of proposing means for better integrating marine research with maritime research. To achieve this, the EU increases the integration between established research disciplines and improves cooperation between all the stakeholders concerned with seas and oceans. This book focuses on results of 13 projects (This work has received funding from the European Union's 7th Framework Program under grant agreement No. 288710 and No. 288192, from the European Union's Horizon 2020 research and innovation programme under grant agreement No. 675680 and No. 773782, from the European Union's INTERREG Balkan-Mediterranean programme under grant agreement MIS 5017160 and from the European Union's European Institute of Innovation and Technology under grant agreement No. 190880, No. 200805, No. 201166, No. 190836, No. 190744, No. 200068 and No. 200620.) funded by the European Commission. These projects propose concrete measures and mechanisms to improve the efficiency and excellence of marine and maritime research in order to address the challenges and opportunities presented by the oceans and seas. This opening chapter provides an

P. Koundouri (✉)
School of Economics and ReSEES Laboratory, Athens University of Economics and Business, Athens, Greece

United Nations Sustainable Development Solutions Network Europe, Paris, France

EIT Climate–KIC Hub Greece, ATHENA Research and Innovation Center, Athens, Greece

ICRE8: International Centre for Research on the Environment and the Economy, Athens, Greece
e-mail: pkoundouri@aueb.gr

V. Manoussi
ICRE8: International Centre for Research on the Environment and the Economy, Athens, Greece

L. Papadaki
United Nations Sustainable Development Solutions Network Greece, Athens, Greece

EIT Climate–KIC Hub Greece, ATHENA Research and Innovation Center, Athens, Greece

P. Koundouri (ed.), *The Ocean of Tomorrow*, Environment & Policy 57,
https://doi.org/10.1007/978-3-030-56847-4_1

introduction to these projects by first reviewing the goals, partners, methodology and objectives of each of the projects.

Keywords Marine ecosystems · Maritime activities · Sustainable oceans · Deep Demonstration · Multi-use offshore platforms · Marine protected areas

1.1 Introduction

Seas and oceans affect our daily lives, providing an essential part of our wealth and well-being. They not only are a critical source of food, energy and resources but also provide the majority of Europe's trade routes. Ocean supplies fresh water (most rain comes from the ocean) and nearly all Earth's oxygen and moderates the Earth's climate, influences our weather and affects human health. It provides foods, medicines and mineral and energy resources. Moreover, it supports jobs and national economies, serves as a highway for the transportation of goods and people and plays a role in national security. A big part of the world's population lives in coastal areas, which are susceptible to natural hazards (tsunamis, hurricanes, cyclones, sea-level change, and storm surges) (Costanza et al. 1999).

Humans affect the ocean in a variety of ways (Fig. 1.1). Laws, regulations and resource management affect what is taken out and put into the ocean. Human development and activity lead to pollution (point source, non-point source and noise pollution), changes to ocean chemistry (ocean acidification) and physical modifications (changes to beaches, shores and rivers). Changes in ocean temperature and pH due to human activities can affect the survival of some organisms and impact biological diversity (coral bleaching due to increased temperature and inhibition of shell formation due to ocean acidification).

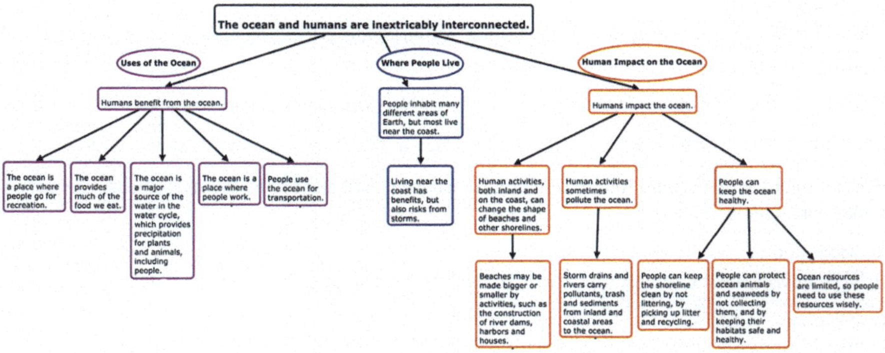

Fig. 1.1 Fundamental interconnection between humans and oceans. (Source: Ocean Literacy Network (2015))

Everyone is responsible for caring for the ocean, and humans must live in ways that sustain the ocean. Individual and collective actions are needed to effectively manage ocean resources for all.

The growing demand for maritime transport, offshore energy, tourism, coastal development, resource extraction, fisheries and aquaculture may have an impact on the marine environment. The United Nations Conference on Trade and Development is projecting an annual average growth rate of 3.4% for the maritime trade for the period 2019–2024 (UNCTAD 2019). The European Union has taken up this challenge and established an integrated maritime policy and highlights the importance of integration between marine and maritime research disciplines, to reinforce excellence in science and to reconcile the growth of sea-based activities with environmental sustainability (European Commission 2017d). The integrated maritime policy covers five cross-cutting policies, namely, Blue growth, which is composed of three components (sustainable jobs and growth, legal certainty and security and cooperation between countries); Marine data and knowledge; Maritime spatial planning; Integrated maritime surveillance; and Sea basin regional strategies.

Another growing challenge for both the marine and maritime sector is climate change. International maritime transport faces a dual challenge in respect of climate change: the need to reduce its carbon emissions and, at the same time, adapt to the potentially wide-ranging impacts of climatic changes (UNCTAD 2020). Ports and shipping are intrinsically linked – as such, efforts to reduce maritime emissions need to extend beyond seagoing ships alone. IMO's MARPOL Annex VI (2010) regulations on air pollution and energy efficiency are aimed at ships. Nonetheless, it is clear that for port emissions to be reduced, emissions from all port-related emission sources need to be addressed. Besides, climate change is expected to have severe impacts on the marine environment. Increase in water temperatures will contribute to a restructuring of marine ecosystems with implications for ocean circulation, biogeochemical cycling and marine biodiversity.

The IPCC (2018) report explicitly refers to the need for "rapid far-reaching and unprecedented changes in all aspects of society". European Union has put into force several directives and regulations aiming to incentivize port and shipping companies to commit to comply with environmental standards. The European Green Deal, the most ambitious action plan of European Union, aims at increasing the EU's greenhouse gas emission reductions target for 2030 to at least 50% compared with 1990 levels. Measures accompanied with an initial roadmap of fundamental policies range from ambitiously cutting emissions, to investing in cutting-edge research and innovation, to preserving Europe's natural environment. (European Commission 2019a).

The main goal of this work has been the contribution to an interdisciplinary and participatory framework of analysis of technical, environmental, economic and social aspects of Blue Growth, which can provide policy recommendations for improved implementation of the Initiative, allied with the relevant Sustainable Development Goals (SDGs), the Marine Strategy Framework Directive (MSFD 2008), the Marine Spatial Planning (MSP) and the Maritime Spatial Planning.

The multidisciplinary team of Professor Phoebe Koundouri joined forced with two well-established networks: UN SDSN and EIT Climate-KIC aiming to drive the highly needed change towards a resilient, circular, net-zero carbon and inclusive economy. EIT Climate-KIC is a European knowledge and innovation community, working towards a prosperous, inclusive, climate-resilient society founded on a circular, zero-carbon economy. The EIT Climate-KIC is part of the European Institute of Innovation and Technology (EIT) and the EIT Community, which is a body of the European Union and a global innovation leader, delivering world-class solutions to societal problems. It seeks to catalyse the rapid innovation needed across sectors by bringing together the brightest minds to tackle challenges, empowering leaders through capacity building, and financially support the most promising climate-positive businesses (EIT Climate-KIC 2020a).

The UN Sustainable Development Solutions Network (SDSN) was set up in 2012 under the auspices of the UN Secretary-General. SDSN mobilizes global scientific and technological expertise to promote practical solutions for sustainable development, including the implementation of the Sustainable Development Goals (SDGs) and the Paris Climate Agreement. SDSN works closely with United Nations agencies, multilateral financing institutions, the private sector and civil society (UN SDSN 2020).

The Oceans of Tomorrow call triggered the interest to explore the whole spectrum of the sustainability concept in the oceans, with several European funded projects having been developed in the last 10 years aiming to respond to the challenges, which oceans face today. The aim of this chapter is to present ten projects that have been selected from Professor Koundouri scientific portfolio aiming to cover challenges and solutions related to sustainable oceans in a holistic manner. The objective is to build the knowledge base for a sustainable growth of sea-based activities. It will do this in three ways: by improving understanding of marine ecosystems' response to a combination of natural and anthropogenic factors, by interpreting the interdependencies between shipping sector and ports and by providing a scientific foundation for feasible, sustainable management and regulating policies taking into consideration relevant innovative technologies.

1.2 The Oceans of Tomorrow

The aims of the call are to improve our understanding and the predictive capacity of marine ecosystems' response to a combination of natural and anthropogenic factors while fostering innovations to make the most of sea resources. The call consists of four topics. Two are of generic nature, multi-use offshore platforms (MUOPs) and marine microbial diversity, while the other two are of particular relevance to the Mediterranean and the Black Sea regions: (1) natural and human-made pressures in the Mediterranean and Black Sea and (2) marine protected areas (MPAs) and wind energy potential in the Mediterranean and Black Sea.

1.2.1 The MERMAID Project

The MERMAID (Innovative multipurpose offshore platforms: planning, design and operation) FP7 European Commission project develops concepts for next-generation offshore platforms for multi-use of ocean space for energy extraction, aquaculture and platform-related transport (MERMAID 2015). The total budget is 7.4 million euro, and the European Union has granted a financial contribution of 5.5 million euro for the duration of 48 months. The project examines different concepts, such as a combination of structures or completely new structures on representative sites under different conditions. In addition, project development guidelines are produced for stakeholders and end users, addressing a wide range of issues such as business and technical aspects and spatial socio-economic planning. The MERMAID consortium consists of 28 European partners from Denmark (Coordinator), Germany, Belgium, Italy, Sweden, Spain, Greece, Norway, the Netherlands, Poland, Turkey, Cyprus and the United Kingdom bringing together expertise from both science and industry (universities (11), research institutes (8), industries (5) and small and medium enterprises (4 SMEs)), from many regions in EU. The project considers four offshore study sites for multi-use offshore platforms. Site-specific designs are being developed based on an extensive stakeholder consultation process and the environmental characteristics of each site. MERMAID focuses on specific challenges in the Baltic Sea, representing a typical estuarine area with fresh water from rivers and salt water from the North Sea; the trans-boundary area of the North Sea-Wadden Sea, representing a typical active morphology site; the Atlantic Ocean, representing a typical deepwater site; and the Mediterranean Sea, representing a typical sheltered deepwater site (Fig. 1.2).

The objectives of the project concerning the effective management procedures are to address the variations in legislation and policies (institutional acceptance), attract developers and investors (financial feasibility) and involve the stakeholders (socio-economic and ecological acceptance). The MERMAID project also aims at the development of innovative technology and design through the development of (a) integrated concepts for extraction of renewable ocean energy, (b) offshore aquaculture technology, (c) large-scale platform design concepts and (d) unification of technologies and services. For the achievement of sustainable integration, the project developed a framework of dynamic and spatial environmental and ecological sustainability and socio-economic viability of multi-use platforms (MUPs). Moreover, for the integration of management, technology and social economics at the four contrasting test sites, the target is to provide tools, techniques and decision support systems that may be applied, tested and validated and are suitable for immediate use. Finally, test sites are used to ensure that MERMAID deliverables are of real value, practicable and usable (MERMAID 2019).

The project consists of nine work packages (WPs): WP1, Project Management; WP2, Assessment of the Policy, Planning and Management Strategies; WP3, Development of Renewable Energy Conversion from Wind and Waves; WP4, Systems for Sustainable Aquaculture and Ecologically Based Design; WP5, Interaction of

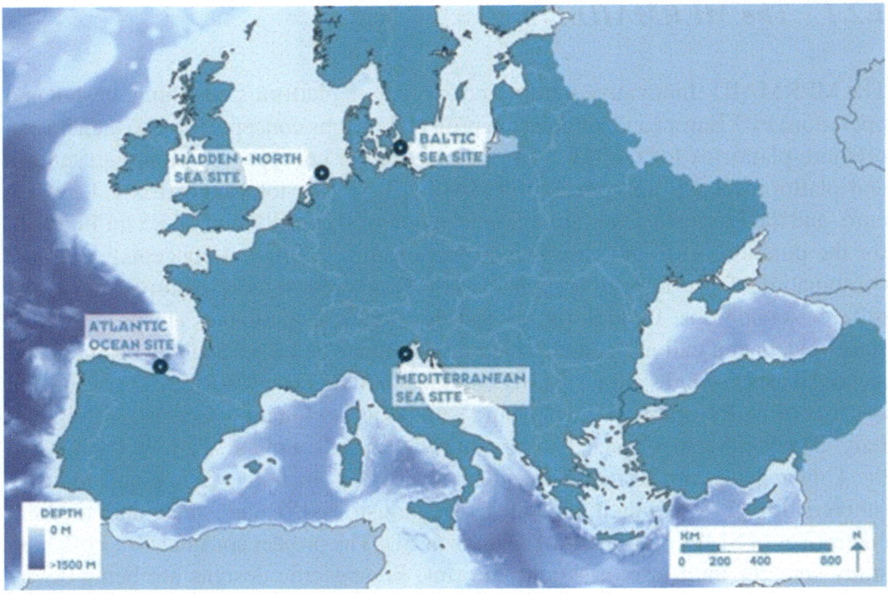

Fig. 1.2 MERMAID sites map. (Source: MERMAID 2015)

Platforms with Hydrodynamic Conditions and Seabed; WP6, Transport and Optimization of Installation, Operation and Maintenance; WP7, Innovative Platform Plan and Design; WP8, Economical, Technical and Environmental Feasibility of Multi-use Platforms; and, finally, WP9, Project Dissemination and Outreach Activities (Fig. 1.3).

The structure of the packages serves the main goal of the project for the development of specific guidelines to assist future stakeholders within the offshore industries in order to plan, establish and operate their businesses in the most optimal way possible.

The final goal of MERMAID is to disseminate project knowledge and procedures into the community of the professionals involved in the planning, installation, operation, maintenance, monitoring and decommissioning of offshore platforms and the preparation and implementation of policies and strategies for sustainable use of sea resources, low carbon economy and eco-friendly transportation. Moreover, specific activities are also planned for the public and businesses which receive benefits from the sustainable development of the use of marine resources, both energy and food. The main dissemination activities include:

(i) The participation in international exhibitions: production of materials, e.g. posters, leaflets, etc.
(ii) The establishment and updating of the project home page on the World Wide Web.
(iii) The publication of peer-reviewed scientific papers.

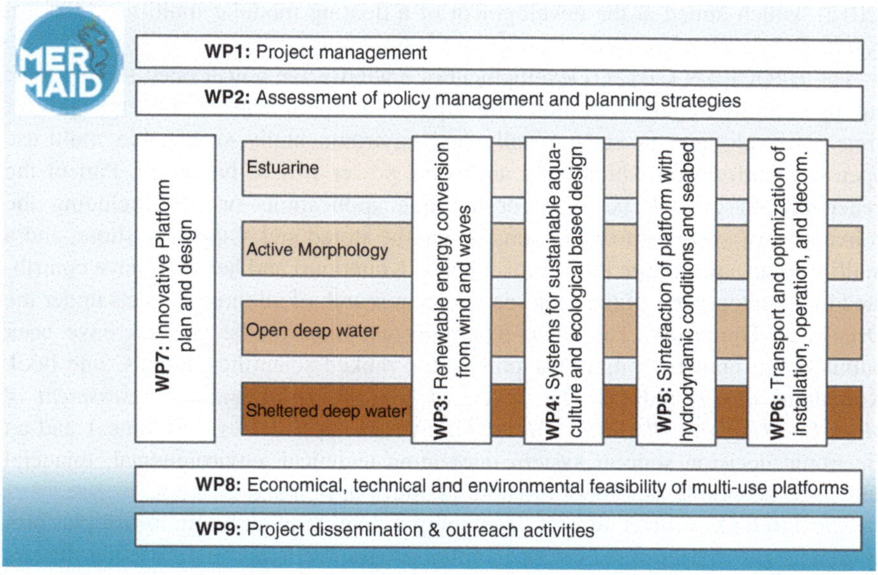

Fig. 1.3 MERMAID work packages breakdown. (Source: MERMAID 2015)

(iv) The active participation of partners in international conferences and publications in the proceedings of these conferences.
(v) The publication of guidelines, based upon the project findings, that present performance and design of innovative multipurpose platforms.
(vi) The utilization of results of MERMAID which are expected to be used in teaching assignments, lectures and exercises at European universities so that the latest state of the art is given to European students.
(vii) The utilization of results in PhD studies in scientific publications and thesis.

Through these activities the MERMAID will accomplish to establish an efficacious collaboration and exchange of knowledge, information and experiences between the scientific community, coastal and marine authorities, the industry and potential end users and also to inform the public about the concept of multi-use offshore platforms.

1.2.2 *The H2OCEAN and TROPOS Project*

In addition to MERMAID, the European Commission granted two other research projects under the call Oceans of Tomorrow with regard to multi-use offshore platforms, allocating a total budget of 20 million; namely the H2OCEAN (2012), which aimed at the development of a wind-wave power open-sea platform equipped for hydrogen generation with support for multiple users of energy, and the TROPOS

(2012), which aimed at the development of a floating modular multi-use platform system for use in deep waters (European Commission 2014a).

The H2OCEAN (2012) (Development of a wind-wave power open-sea platform equipped for hydrogen generation with support for multiple users of energy) project aims at developing an economically and environmentally sustainable multi-use open-sea platform on which wind and wave power will be harvested. Part of the generated energy will be used for multiple applications on-site, including the conversion of energy into hydrogen that can be stored and shipped to shore, and a multi-trophic aquaculture farm. Prof. Phoebe Koundouri and her team have contributed to the integration of the socio-economic research of all three projects under the Oceans of Tomorrow. The socio-economic outcomes of the projects have been published in numerous multidisciplinary top-ranked scientific journals, one book Koundouri (2017) (editor) *The Ocean of Tomorrow: Investment Assessment of Multi-Use Offshore Platforms: Methodology and Applications* - Volume 1 and an electronic decision support system integrating technical, environmental, financial and socio-economic aspects of offshore platform design and location.

The TROPOS project aims at developing a floating modular multi-use platform system for use in deep waters, with an initial geographic focus on the Mediterranean, tropical and subtropical regions, in particular on the EU outermost regions (OMRs), composed of the Azores, the Canary Islands, Guadeloupe, Guiana, Madeira, Martinique and Reunion, but designed to be flexible enough so as not to be limited in geographic scope. TROPOS gathers 20 partners from 9 countries (Spain, the United Kingdom, Germany, Portugal, France, Norway, Denmark, Greece and Taiwan), under the coordination of PLOCAN (Spain). The TROPOS project is a €7-million European project aiming to explore the design of offshore multi-use platforms, in which a mixture of different sectors and specific functions can be performed in a shared location with shared infrastructure (and costs) and could prove to be an important opportunity for improved utilization of the oceans as well as sustainable economic growth (Fig. 1.4). Developing a concept of multi-use oceanic platforms has become one of the EU's most interesting bets to guarantee the use and synergistic exploitation of oceanic resources in a sustainable and eco-friendly manner.

The TROPOS project enhances the potential and increases the added value of the integration of the four disciplines of Transport, Energy, Aquaculture and Leisure (TEAL) in the proposed floating platform designs. This is to be achieved by ensuring a maximum level of integration and synergy of the multiple platform uses and studying and proposing how to enhance the potential of TEAL integration. The main objectives of the project are to (i) determine suitable locations for the offshore platform system and explore the relations, potential for synergies and integration of a broad range of functions including different marine renewable energy sources, fisheries, aquaculture and related maritime transport aspects; (ii) develop innovative designs of MUPs that allow their close spatial integration and study logistics, security, installation, operational, decommissioning and maintenance requirements for MUPs; and (iii) study the economic viability, socio-economic acceptability and environmental impact of key combinations and address ways to minimize significant

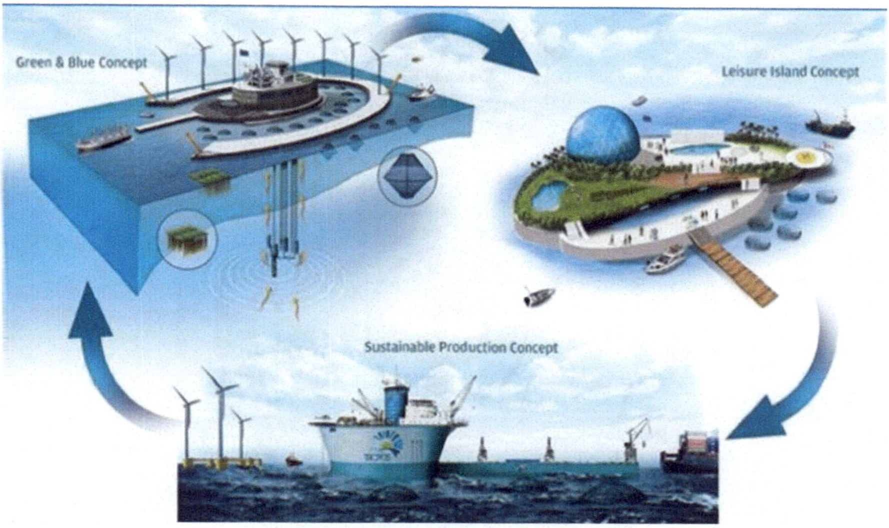

Fig. 1.4 The TROPOS project. (Source: TROPOS 2012)

negative impacts of floating MUPs and configure the MUPs for the Mediterranean, subtropical and tropical areas.

TROPOS project is divided into eight work packages (WPs), which are completely interdependent and interactive with one another: WP1, Project Management; WP2, Geographic and Module Benchmarking and Decision Methodology; WP3, Conceptual Design of Platform Components and Integration; WP4, Engineering Specification for Chosen Platform Designs; WP5, Strategy: Economics, Infrastructure and Logistics; WP6, Environmental and Socio-Economic Impact, Legal Issues; WP7, Communication, Dissemination and Technology Transfer; and WP8, Scientific and Technical Coordination. The successful completion of all the work packages (Fig. 1.5) allows reaching the planned objectives of the project.

Throughout the project, the TROPOS consortium will be disseminating the results of the project to multiple audiences, via different channels, at different intervals, etc. This is necessary throughout the project to raise awareness and invite debate and comment on the scientific work which is being undertaken while it is in progress as well as for the dissemination of the project's deliverables and results to promote uptake and increase impact of the project and its visibility and credibility as possible. The project's dissemination strategy uses the Internet, presentations, workshops, publications and the online project site. The open access project deliverables, results, documentation, scientific publications and newsletters ensure the awareness of the public.

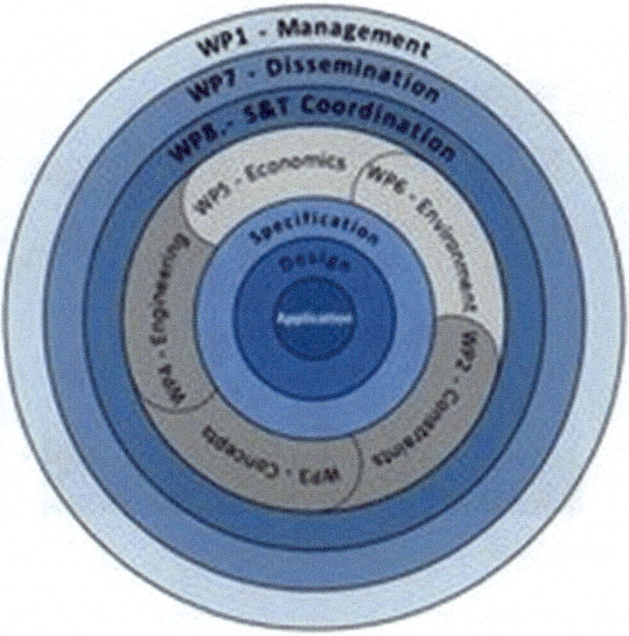

Fig. 1.5 TROPOS work packages breakdown. (Source: TROPOS 2012)

1.3 Blue Growth and Maritime Spatial Planning

The European Commission has initiated a programme called Blue Growth, which is the EU's long-term strategy to support sustainable growth in the marine and maritime sectors. Marine space will become a delicate issue to future Blue Growth plans in European seas, challenging creative and innovative contributions which at the same time can ensure environmental sustainability. One such creative and innovative possibility is multi-use, i.e. making use of marine space more efficiently and effectively when two or more sectors join activities in same area (European Commission 2017a). Whereas this is not fully new, as we used to combine, for instance, shipping and fishing, the approach is now expanded to cover typical Blue Growth sectors, such as wind and aquaculture combinations. While the multi-use arrangement increasingly is accepted as a valuable contribution to Blue Growth, it is not clear how to practically implement it in European seas.

1.3.1 The THAL-CHOR Project

THAL-CHOR (2015) (Cross-border cooperation for Maritime Spatial Planning Development) Interreg project seeks the development of a methodology for

Maritime Spatial Planning and then the adoption of this methodology for pilot application in selected areas in Cyprus (Limassol area) and Greece (Islands of Lesvos and Rhodes). The resolution of spatial conflicts of the sea, better coordination between stakeholders and strengthening cross-border cooperation were also the project's objectives. The consortium is composed of six partners from Cyprus and Greece including the Ministry of Transport, Communication and Works (CY), who is the lead partner, the Ministry of Interior (CY) and the Ministry of Shipping and Island Policy (EL).

The work packages were composed of the following actions, namely, analysis of main features of the marine environment and its benefits for human activities; development of a WebGIS to display all collected data; overview of the legal framework and recommendations for its improvement; definition of future priorities and analysis of the future state in terms of evolution of existing activities and development of new ones; pilot implementation of Maritime Spatial Planning in selected areas and drafting of pilot maritime spatial plans; and evaluation of the methodology followed for implementing Maritime Spatial Planning and identification of good practices.

1.3.2 *The BlueBRIDGE Project*

BlueBRIDGE (Building Research environments fostering Innovation, Decision-making, Governance and Education to support Blue Growth) H2020 project supports capacity building in interdisciplinary research communities actively involved in increasing scientific knowledge about resource overexploitation, degraded environment and ecosystem with the aim of providing a more solid ground for informed advice to competent authorities and to enlarge the spectrum of growth opportunities as addressed by the Blue Growth Societal Challenge (BlueBRIDGE 2020). BlueBRIDGE bundles forces from intergovernmental organizations, research institutes, industry and SMEs establishing a network with a proven track in VREs and e-infrastructures, computer science, marine, aquaculture, environmental and fisheries science and economy.

The project consists of ten WPs: WP1, Project Management; WP2, Project Governance, Exploitation and Sustainability; WP3, Communication, Stakeholder Engagement and Knowledge transfer; WP4, VREs Deployment and Operation; WP5, Supporting Blue Assessment: VREs Development (FAO, Anton Ellenbroek), WP6, Supporting Blue Economy: VREs Development; WP7, Supporting Blue Environment: VREs Development; WP8, Supporting Blue Skills: VREs Development; WP9, VRE Commons Development; and WP10, Interfacing Infrastructures.

BlueBRIDGE has developed innovative services, the so-called Virtual Research Environments (VREs), in four major areas: Fisheries through an ecosystem approach, where services for stock assessment and for the generation of unique identifiers for global stocks are developed; Aquaculture, where services supporting the analysis of socio-economic performance in aquaculture are provided (Tsani and

Koundouri 2018); Maritime Spatial Planning, where spatial planning services are available to identify aquaculture and fisheries infrastructures from satellite imagery and tools to visualize, analyse and report on a range of ecologically important seafloor features within marine protected areas (MPAs); and Education, where tools to set up and deliver training courses in a cost-effective way are developed.

1.4 Cultural Ecosystem Services and Marine Protected Areas Management

Marine protected areas (MPAs) are vital for the conservation of the Mediterranean Sea, due to their role in managing several pressures like fishery. Nonetheless, many species and habitats in MPAs are still exposed to a variety of stressors. In most MPAs human activities are not spatially managed considering their cumulative effects, while monitoring and protection are not coordinated. Additionally, limited effort has been carried out to address the environmental challenges, to measure the consequences of human impacts and to provide management recommendations.

1.4.1 The AMAre Project

The AMAre (Actions for Marine Protected Areas) Interreg project aims to improve the management and protection measures in order to maintain the biodiversity and to increase the resilience of the MPAs for current purposes and upcoming challenges, strengthening the sustainable use of the resources. The AMAre consortium consists of ten European partners from Italy (coordinator), Malta, Spain, Greece and France with the pilot activities being developed in four MPAs, tailored on the specific human uses. The pilot cases will be in Spain (the Freus d'Eivissa i Formentera Marine Reserve), Malta (Żona fil-Baħar bejn Il-Ponta ta' San Dimitri (Għawdex) u Il-Qaliet), Italy (Porto Cesareo MPA and Torre Guaceto MPA) and Greece (National Marine Park of Alonissos Northern Sporades).

MPAs are exposed to a variety of stressors, related to human activities. Unsustainable use of resources leading to habitat loss, lack of knowledge of the mismanagement economic loss, lack of coordination and monitoring of biological and environmental changes in MPAs and lack of integration of the interdependencies amongst human activities and MPAs in the future (e.g. coastal tourism, aquaculture, offshore wind farms) are some of the reasons why environmental accounting of the MPAs is needed.

The focus of the project was concentrated on those human uses considered a concrete threat (fishery, touristic frequentation, yachting) including those possibly increasing in the future (e.g. marine litter), which are difficult to be managed both at local and at transnational scale. The pilot activities are replicated by different partners using the same scheme but considering different preferences to ensure great applicability of the project results. Local and regional stakeholders (e.g. management bodies of protected areas, MPA managers) will benefit from these pilot activities since they consist the first concrete example of coordinated application of marine spatial planning (MSP) in the Mediterranean Sea able to significantly improve the management and the monitoring of biodiversity and the ecosystem services provided by MPAs.

Its main objectives are the refinement of a general framework, the development of shared methodologies (spatial planning and monitoring) and geospatial tools for multiple stressors assessment, the environmental monitoring and stakeholders engagement seeking to structure and exchange information amongst MPAs on biodiversity status, environmental variables, distribution and intensity of human pressures focusing on vulnerable habitats of EU importance (e.g. coralligenous outcrops and *Posidonia* meadows) (AMAre 2020).

Amongst the project priorities were: the redistribution of human activities reducing conflicts and increasing synergies; the implementation of well-designed monitoring activities shared and comparable across MPAs and habitats, allowing the objective to comparatively assess the performance of single MPAs versus an MPA network; the investigation of the potential to individuate early warning indicators of changes; and the development of common trans-frontier regulations and best practices to deal with present and future drivers of changes. The final aim of AMAre is to scale up strategies and recommendations at transnational level adopting an ecosystem-based approach considering the goals of the Marine Strategy Framework Directive (MSFD 2008) across MPAs.

The AMAre WebGIS (https://amare.interreg-med.eu/toolbox/geoportal/) created by the project is a web-based portal for interactive visualization of the spatial data collected in the study areas, organized in a common spatial infrastructure. The portal combines data relevant for the management of the MPA such as administrative, biodiversity, elevation, geology, habitats and biotopes, hydrography, monitoring, oceanography and socio-economic threats.

1.4.2 The RECONNECT Project

The RECONNECT (Regional cooperation for the transnational ecosystem sustainable development) Interreg project studies four marine protected areas and Natura 2000 sites, each of them belonging to a different country, meaning Greece, Cyprus, Bulgaria and Albania (RECONNECT 2020). The NATURA 2000 is the ecological network for the conservation of wild animals and plant species and natural habitats of community importance within the Union. It consists of sites classified under the Birds Directive and the Habitats Directive (the Nature Directives). On the other hand, the MPAs are areas designated to protect marine ecosystems, processes, habitats and species, which can contribute to the restoration and replenishment of resources for social, economic and cultural enrichment. The targeted ecosystems are understudied in the Eastern Mediterranean and the Black Sea regarding the rest of

Europe, and many of the field measurements and estimations are performed for the first time in all study areas. Let aside their economic evaluation with the most state-of-the-art socio-economic methodology. The project's overall objective is the development of management scenarios which can be of use for the policymakers to make informed decisions that can maximize the society's well-being based on the services they receive from those MPAs.

Essential Socio-economic and Cultural Variables offered by the *Posidonia* oceanic habitats were identified in the regional marine park of northern Karpathos Saria and Astakidonision (Greece); the MPA of Kavo Gkreko (Cyprus); the protected area of Ksamil Bay and islands – Stillo Cape – Tongo Island (Albania); and the *Zostera marina* in the Natura 2000 of Gradina-Zlatna Ribka (Bulgaria).

The four study areas face similar threats and pressures based on underwater field studies. Their current ecological status necessitated a choice of Ecosystem Services (ES) to be researched that are both sites specific in order to provide the local policymakers with information for decisions able to maximize local well-being and important at a regional level in order to provide them with conclusions regarding regional development.

In RECONNECT the use of Choice Experiment (CE) was considered the most state-of-the-art method for the ES from these specific species (Vassilopoulos and Koundouri 2017). The ES evaluated are fish abundance, sea water clarity, aesthetic benefits, carbon sequestration, protection from erosion and preservation of underwater cultural heritage. An extra aspect was included as the most common pressure in all areas which was the overfishing, while in the case of Cyprus, the impact from massive tourism was incorporated as well.

The project's innovation relies on the expansion of the existing socio-economic technique for the evaluation of a category of ES that is not studied to date due to the lack of standardized methodology and its difficulty to be captured. This is the Cultural Ecosystem Services (CES) referring to the physical and intellectual interactions, the sense of belonging or collective identity that people experience when they snorkel or dive in areas with *Posidonia* seabed. It is also the ocean literacy, research and promotion of education on the marine environment that accrues from its conservation and protection. The two CES studied in RECONNECT are the "aesthetic benefits" related to the enjoyment of sea wildlife that offers a sense of unimpacted nature when people come in contact with the iconic or non-iconic species hosted in *Posidonia* habitats with some of them being emblematic (f.e. monk seals) and the protection of underwater cultural heritage related to *Posidonia*'s ability to provide a protective matt above archaeological treasures. It buries deeper sediments and provides the right conditions needed to preserve underwater remains by locking out oxygen that otherwise degrades them.

1.4.3 Cultural Heritage Protection from Anthropogenic Climate Change

Natural and cultural heritage represents a key socio-economic capital and offers many different benefits to citizens. Extreme weather conditions and the adverse effects of climate change are able to significantly damage cultural and natural heritage. If we do not act immediately, the damage may be irreversible. A Greek emblematic initiative for the protection of monuments of cultural heritage from anthropogenic climate change was launched in 2019 at "Climate Change Impacts on Cultural Heritage: Facing the Challenge" international conference. The aim of this initiative is to track the effects of climate change on these monuments, as well as to assess the social and economic values at stake. It will also address critical challenges to be faced by decision-makers, who are managing these monuments.

1.5 Sustainable Coastal and Rural Areas

1.5.1 The COASTAL Project

Rural development in the EU is affected by several environmental, economic and social pressures, such as changing market developments, decreasing population densities, lack of employment, desertification and others. However, coastal areas provide appealing business opportunities while they are influenced by economic activities in the hinterland (European Commission 2017b). To analyse the environmental, economic and social interactions of rural and coastal areas, multi-actor approaches need to be combined with System Dynamics in a holistic manner.

The COASTAL (Collaborative Land-Sea Integration Platform) H2020 project aims to improve the rural-coastal synergies in strategic business and policy decision-making and collaboration between coastal and rural actors. A generic toolset and performance indicators are being developed, demonstrated and applied by combining a multi-actor approach with system dynamics modelling, which enables the understanding of the interactions with market, demographic, environmental and climate forecasts and the quantification of the positive and negative externalities (COASTAL 2020).

COASTAL is a unique collaboration of 29 partners from 8 EU Member States, representing coastal and rural business entrepreneurs, administrations and scientific experts. The project core consortium includes 11 research institutes and 3 universities active in the field of marine science and innovation, hydrology, rural development, agriculture and integrated systems modelling; 3 NGOs active in the field of regional development and economics, agriculture, tourism and coastal development; 2 farming advisory organizations; 4 administrations involved in regional and rural development, port development and environmental management; 2 SMEs with expertise

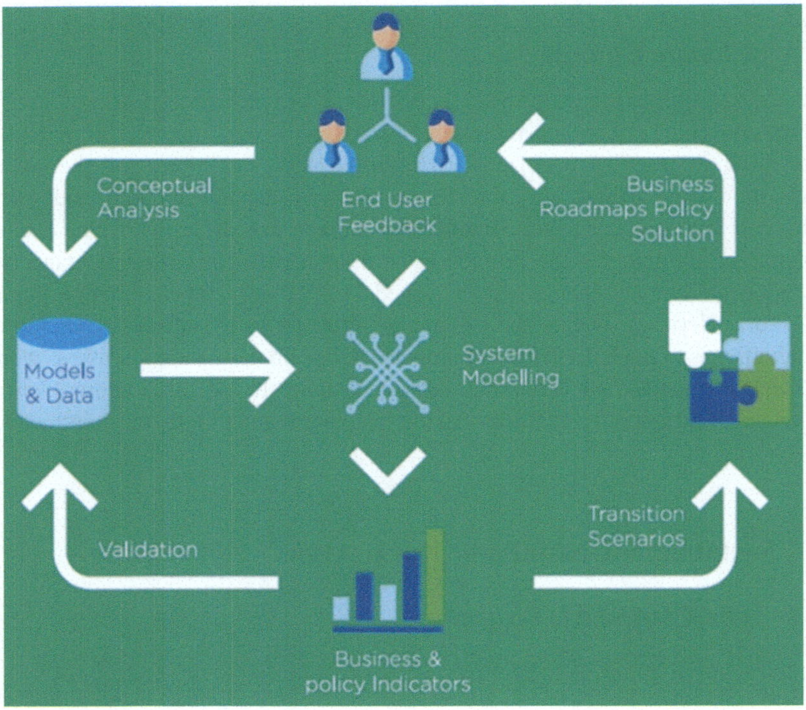

Fig. 1.6 The COASTAL process. (Source: COASTAL (H2020))

in knowledge dissemination, blue growth and industrial and coastal development; and 2 development agencies and 2 partners representing the business sector.

The process followed in COASTAL (2020) projects can be grouped into the following steps, as presented in Fig. 1.6. Local actors and experts participate in collaborative exercises to go through the underlying causes, propose and discuss solutions and validate and interpret the impacts of simulated business and policy decisions. Qualitative and quantitative techniques are combined in this co-creation process supported by graphical tools to gain an in-depth understanding of the systemic transitions underlying the land-sea interactions. These systemic transitions are synthesized and analysed with dynamic models to produce multiple transition scenarios for core business and policy indicators. From these, practical business road maps and policy solutions are derived, which are easily integrated in the models used to support the analyses.

In a nutshell, the project seeks to contribute to the long-term improvement of sea water quality while creating added value and jobs in coastal areas and the hinterland; develop a transferable set of tools and indicators allowing qualitative and quantitative analysis of land-sea interactions for evidence-based policymaking; and provide thorough understanding of the barriers and motivators affecting behaviour and solutions enabling joint actions. Meanwhile, it aims at increasing the job potential and creating added value in coastal areas resulting from new business opportunities and improved collaboration between land- and sea-based operators; reducing externalities from land-based on sea-based activities by improving economic collaboration and integrated governance; and creating durable relationships between coastal areas, serving as flagships for coastal-rural synergy.

1.5.2 *The Marine Plastic Litter Challenge*

The Mediterranean Sea is a top tourism destination in the world hosting nearly 314 million international tourists a year, with European Mediterranean countries attracting most of the tourists, but it is also one of the most affected areas by marine litter worldwide, polluting its shores and pristine coastal waters (UNEPMAP 2020). The plastic litter problem in seas and oceans has several environmental impacts associated. Besides the direct effect on the marine environment, marine littering is often related to socio-economic characteristics degradation (e.g. the food chain unsettling, public health threatening, loss of jobs and property devaluation).

EIT Climate-KIC is a Knowledge and Innovation Community (KIC), working to accelerate the transition to a zero-carbon economy. Supported by the European Institute of Innovation and Technology, EIT Climate-KIC identifies and supports innovation that helps society mitigate and adapt to climate change. They bring together partners in the worlds of business, academia and the public and non-profit sectors to create networks of expertise, through which innovative products, services and systems can be developed, brought to market and scaled up for impact. Through their convening power, EIT Climate-KIC brings together the most effective groups to create the innovation that can lead to systemic change (EIT Climate-KIC 2020a).

WEF (2016) reports that the circular economy is gaining growing attention as a potential way for our society to increase prosperity while reducing demands on finite raw materials and minimizing negative externalities. Such a transition requires a systemic approach, which entails moving beyond incremental improvements to the existing model as well as developing new collaboration mechanisms. BL.EU Climate (Climate Innovation in Southern Waters) funded by EIT Climate-KIC BL.EU seeks to address the challenge of plastic marine littering in southern European waters by building capacity for innovation to address the issue at the very beginning of its life cycle, on the prevention side and plastic waste reduction with significant climate change mitigation potential from the reduction in the collected and handled plastic waste. Greece, Portugal and Croatia gathered around this problem and identified three pillars around ports (commerce, fishing, tourism) working closely with local problem owners: in Croatia, islands of Cres and Zlarin; in Greece, the port of Piraeus and islands of Milos and Andros; and in Portugal, the port of Lisbon (BL.EU. Climate 2020).

The project enhances plastic litter reduction through a systems innovation approach, where stakeholders are involved through participatory workshops in solutions identification. Stakeholder mapping, validation interviews and surveys were performed targeting tourists and fishermen only. The results of the questionnaires were analysed and presented at different workshops conducted in the project sites. The main objective of the workshops was to trigger a discussion amongst the participants (mostly stakeholders identified at the mapping exercise) on potential solutions to prevent, reduce and collect marine litter, focusing on plastics. All the above led to the design of a strategic road map by all three countries, identifying steps to reduce the negative effects caused by plastic waste in the future, supporting not only governments but also regions, municipalities, industries, consumers and civil society to improve the awareness campaigns, systems design, replacement, refuse, recycling and reuse of plastic.

BL.EU Climate project is followed by another project supported by EIT Climate-KIC, MEDfreeSUP (tackling single-use-plastic item uses in the Eastern Mediterranean Sea), which is based on the plastic waste prevention (PWP) approach and aims to enable local ecosystems to move towards reusable materials. The project focuses on the East Mediterranean coast, targeting the three biggest coastal countries: Italy, Croatia and Greece. The main objective of the project is to set a replicable voluntary protocol for free single-use plastics food packaging adoption for cafes, restaurants, foods stores, hotel and beach facilities but also for public events and places to provide support and guidance to local business in order to comply to the EU SUP Directive and go beyond the law to engage Mediterranean islands and cities in the transition towards a free single-use plastic environment.

1.6 Sustainable Shipping and Ports

Sustainable shipping refers to the broad set of challenges, governance rules and regulations, management patterns and corporate behaviours, stakeholders' engagement and industrial activity forms that should come to define a maritime transport industry that is shaped by the broader societal goals of sustainable development. The port industry together with the shipping industry constitutes a key node in the international supply chain considering that over 80% of volume (70% of value) of world's merchandize trade is carried by sea (UNCTAD 2019).

Environmental challenges relating to shipping sector and ports are twofold, namely, the effects of maritime transport on the environment (e.g. pollution, CO2 emissions) and conversely the environmental impact on maritime transport (e.g. water level, floods, storms, precipitation and extreme weather events) (Asariotis et al. 2018). Shipping and ports are intrinsically linked – as such, in order to reduce maritime emissions, joint efforts by ports, shipping companies, regulating authorities and key stakeholders need to be aligned and orchestrated. IMO's MARPOL Annex VI (2010) regulations on air pollution and energy efficiency in alignment with several European regulations and directives (as presented further in Chap. 11) can be a starting point of implementation and behaviour change if addressed appropriately.

Ports role in enforcing the International and European policy and so, in driving the emissions reduction, comes down to adopting measures such as introducing

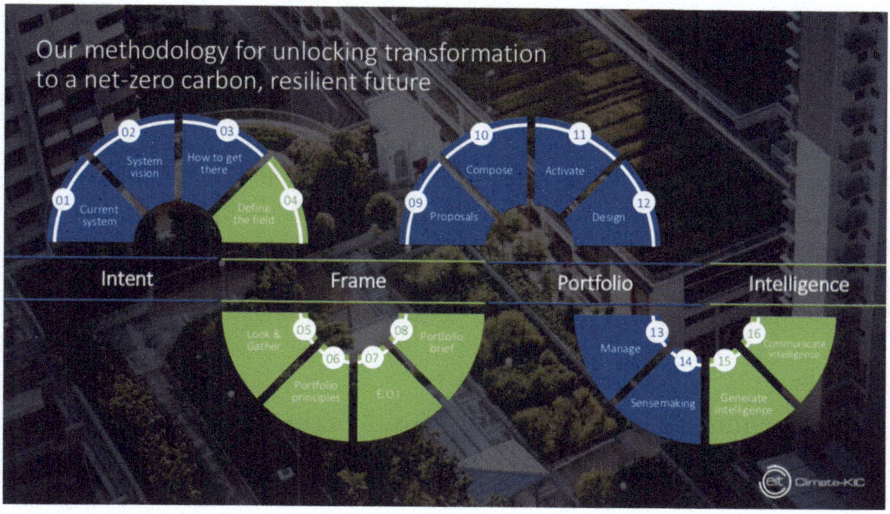

Fig. 1.7 Deep Demonstration methodology. (Source: EIT Climate-KIC (2020))

differentiated port dues, providing onshore power supply (e.g. "cold ironing"), switching to low-sulphur fuels and setting speed limits in ports. Ports are increasingly expected to align their performance with sustainability expectations, i.e. to deliver optimum economic and social gains while causing minimum environmental damage. In view of this challenge, ports and consequently the shipping sector need to understand the interdependences between maritime, marine and urban activities (e g. port areas).

A systematic approach to aligning the expectations and needs of key stakeholders into an environmental management system would enable the continuous identification of an individual port's priorities while it introduces a functional organizational structure that sets respective targets, implements measures, monitors impact, evaluates, reviews and takes corrective actions when and where necessary. In a phased way, EIT Climate-KIC works with a consortium of high-ambition port authorities in Valencia (Spain) and Piraeus (Greece) and Cyprus Ministry of Shipping to demonstrate how ambitious maritime hubs can be catalysts for reversing the fast-growing emissions from international shipping and trade using systems innovation approach.

Deep Demonstrations funded by EIT Climate-KIC start with a demand-led approach, working with organizations willing to take on the responsibility of acting as "problem owners" – in Greece Piraeus Port Authority – committed to zero-net emissions, resilient futures (EIT Climate-KIC 2020b). Deep Demonstrations (Fig. 1.7) progress in tightly designed, iterative phases – steps of rolling out systems innovation-as-a-service, aiming at the identification of the key actors to be involved, current status, vision, innovation needs and sustainable financial planning and ultimately at the alignment of all actors able to drive systems transition to a low-carbon emissions future. Deep Demonstration is a circular approach in innovation implementation with the final goal the holistic change of the port to Sustainability.

The aim of Deep Demonstration in Maritime Hubs is to drive the decarbonization of the three European ports considering all major aspects and conflict points, bringing the key stakeholders in a consortium where impactful discussions take place, co-creating a common vision and working together towards meeting the Sustainability needs of the port through innovation and synergies. Deep Demonstration is part of a broader UN SDSN Initiative, namely, the Global Roundtable for Sustainable Shipping[1] which aims at bringing together shipowners, shipbuilders, technology developers and researchers, ports and policymakers from across the globe. The main goal of the initiative is to develop zero-emission shipping innovations, having as a target net-zero emissions by 2050.

1.7 Circular Economy and Decision-Making

Despite the growing alarms over climate change threat and the need for financial tools and resources to be mobilized, limited action has been noted in most European countries in the last decade. The EU has set an ambitious plan for the adoption and implementation of the circular economy (CE). The Plan estimates that by 2030 the integration of CE will result in savings of over 600 bil. euros for EU businesses, will create 580,000 new jobs and will contribute to the reduction of 450 mil. tons greenhouse gases. Member States are key players in Europe's transition to a circular economy. Three EIT Climate-KIC projects seek the implementation of circular approaches in critical public and private institutions.

Adopting Circular Economy in each country should be aligned with its strategic documents and identified sectoral strengths and needs, set in the individual country's Smart Specialization Strategies. CE in S3 (Circular Economy Transition in Smart Specialization Strategy) project aimed at piloting the adoption of CE in their respective S3s, working together with the responsible authorities in two European countries, Greece and Bulgaria. Specific approaches to implementation and supporting funding instruments were discussed and shared with participants. The end goal of the project was to stimulate the timely and systemic adoption of the CE in S3s for the 2020–2027 programming period in EU Member States.

Circular Learning Hub (a learning hub for the engagement and ecosystem transition towards circular thinking) works on an awareness-intention-action path fostering problem owners to a deeper understanding of circular thinking. Two lab experiments will be performed by the end of 2020 to empirically confirm the impact of ambiguity on decision-maker's (e.g. investors and entrepreneurs) prior probability function and posterior distribution. The experiments will aim to estimate a parametric model of attitudes towards risk/ambiguity and time preferences considering also the impact of climate change (through VR) on decision-making process.

[1] http://unsdsn.gr/global-roundtable-for-sustainable-shipping-2

In a choice experiment, individuals are in a hypothetical setting and asked to choose their preferred alternative amongst several alternatives in a choice set. Each alternative is described by a number of attributes or characteristics. Thus, individuals implicitly make trade-offs between the levels of the attributes in the different alternatives presented in a choice set. Usually, a monetary value is included as one of the attributes when describing the profile of the alternative presented. Virtual Reality Laboratory Experiments are described in three-dimensional topographies containing virtual objects obeying simulated physical laws. The information collected during the experiment will allow for the co-creation of innovative learning initiative on circular thinking. The pan-European extension of this project (Italy, Greece and Bulgaria) will also allow for a more effective co-creation of scaled nudging and learning solutions based on the experiences of the different geographical contexts involved.

The outcomes of Circular Learning Hub will be further exploited by CE Beacons (Western Balkan Circular and Climate Innovation Beacons). The goal of this project is to build an eco-system for circular, climate-related innovation that catalyses systems transition. The project tackles the following objectives, namely, the establishment and empowerment of a network of actors in the Balkans and other countries for developing and implementing circular innovation in business and policy; the creation of partnerships amongst EU countries through shared circular vision, information, resources and experience on both strategic and operational level; the nurture shared learning amongst "Beacons" to raise capacity of places for systems circular change and climate action; and the reduction of the total negative impacts from current market design through establishing circular markets.

1.8 Mobilizing Science-Driven Sustainable Blue Growth

As a response to the climate crisis and its effect on Marine Ecosystems and Coastal Populations, a Euro-Asian initiative has launched by the SDSN Mediterranean and the Black Sea regional networks together with the national SDSN networks in Greece, Italy, Spain, France, Turkey and Russia under the auspices of the Global SDSN network IN 2019. The goal of the 4-Seas Initiative[2] is to accelerate science-driven blue growth and the implementation of the United Nations Sustainable Development Goals (Agenda 2030) in the following four seas: the Mediterranean, the Black, the Caspian and Aral seas. It follows a "source-to-sea" approach initiating from river basins and proceeding to coastal and marine ecosystems, as well as the societies whose livelihoods depend on these ecosystems.

The initiative focuses on five areas, namely, research, education, data management, innovation and policy, while it brings together all relevant stakeholders in co-designing a systems innovation pathway for the transition to socially, culturally,

---

[2]http://unsdsn.gr/un-sdsn-4-seas-initiative-mobilizing-science-drive

economically, environmentally and geopolitically sustainable development in the Euro-Asia Region embracing the 4-seas. Categorically, it seeks the development of Sustainable Blue Growth Transformation Pathways in urban and rural areas that depend on the 4-seas; the creation of networks of protected marine and coastal ecosystems of cultural and natural interest; the creation of educational programmes with an emphasis on understanding and implementing the SDGs while enriching them with Euro-Asian intellectual tradition and the development and support of open access databases with the data, models, results and policy recommendations relevant to the sustainability transition of the 4-seas; the incubation and acceleration of technological and social innovation for the sustainability transition in the region, combining the protection of the natural and cultural environment; and policy recommendations for the support of sustainable blue growth in the region.

1.9 Conclusions

The cross-thematic *Oceans of Tomorrow* book seeks to implement this commitment. The book falls within the activities launched under European calls (e.g. FP7, H2020) to implement the European strategy for marine and maritime research and to address marine sciences and technologies as a challenge that cut across themes. This book aims to foster multidisciplinary approaches and cross-fertilization between various scientific disciplines and economic sectors on key cross-cutting marine and maritime challenges. Research projects presented in the next chapters bring together scientists, technology providers, industrial partners (including Small and Medium Enterprises – SMEs) and end users.

There is no doubt that the Earth's survival will depend on the protection and sustainable management of our seas and oceans and the resources they provide. This is recognized by the Joint Communication on International Ocean Governance, which is an integral part of the EU's response to the United Nations' 2030 Agenda for Sustainable Development, and in particular to the targets set out by Sustainable Development Goal 14 (SDG 14) to "conserve and sustainably use the oceans, seas and marine resources". The EU's Seventh Framework Programme for Research and Development (FP7) and Horizon 2020 have funded over 1200 Blue Economy-related projects (European Commission 2014b, 2020). In addition to the political priorities, the challenges of the protection and sustainable management of our seas and oceans are directly linked to the EU Bioeconomy Strategy and Blue Growth Strategy (European Commission 2017c, 2019b).

References

AMAre. (2020). *Interreg European Commission project. Actions for Marine Protected Areas.* Online Platform. https://amare.interreg-med.eu. Accessed on 18 Mar 2020.

Asariotis R., Benamara, H., & Mohos-Naray, V. (2018). UNCTAD research paper no. 18 UNCTAD/SER.RP/2018/2018/Rev.1 United Nations conference on trade and development (2018) file:///E:/all%20docs%20and%20files/Article%20on%20Sustainable%20ports/Unctad%20survey%20on%20ports_ser-rp-2017d18_en.pdf

BL.EU. Climate. (2020). *EIT Climate-KIC project*. Climate Innovation in Southern Waters. https://www.athenarc.gr/el/climate-innovation-southern-european-waters-bleu-climate

BlueBRIDGE. (2020). *Horizon 2020 European Commission project. Building research environments fostering innovation, decision making, Governance and Education to support Blue growth*. https://www.bluebridge-vres.eu/about-bluebridge. Accessed on 11 Feb 2020.

COASTAL. (2020). *Horizon 2020 European Commission project*. Collaborative Land-Sea Integration Platform. https://h2020-coastal.eu

Costanza, R., et al. (1999). Ecological economics and sustainable governance of the oceans. *Ecological Economics, 31*(2), 171–187.

EIT Climate-KIC. (2020a). *Europe's leading climate innovation initiative*. https://www.climate-kic.org/who-we-are/what-is-climate-kic/

EIT Climate-KIC. (2020b). *Deep demonstrations*. https://www.climate-kic.org/programmes/deep-demonstrations/. Accessed on 12 Apr 2020.

European Commission. (2014a). *The ocean of tomorrow projects (2010–2013)*. https://publications.europa.eu/en/publication-detail/-/publication/95d22319-ce5e-4a60-889b-5620104610b2/language-en/format-PDF/source-search

European Commission. (2014b). *Selected projects in marine research*. https://publications.europa.eu/en/publication-detail/-/publication/b522116e-af4d-427f-8c15-92d732c2595c/language-en/format-PDF/source-71778581

European Commission. (2017a). *Sustainable blue economy*. https://publications.europa.eu/en/publication-detail/-/publication/ada65c0f-aef9-11e7-837e-01aa75ed71a1/language-en/format-PDF/source-71790365

European Commission. (2017b). *Common organization of agricultural markets (COM)*. https://publications.europa.eu/en/publication-detail/-/publication/5e3053c2-d6cb-11e7-a506-01aa75ed71a1/language-en/format-HTML/source-72100476

European Commission. (2017c). *Blue Growth – Shaping the next five years together*. https://ec.europa.eu/maritimeaffairs/content/blue-growth—-shaping-next-five-years-together_en

European Commission. (2017d). *Integrated maritime policy*. https://ec.europa.eu/maritimeaffairs/policy_en

European Commission. (2019a). *The European Green Deal, Brussels, 11.12.2019 COM*. Final Communication from the commission to the European Parliament, the European Council, the council, the European Economic and Social committee and the committee of the regions. https://ec.europa.eu/info/sites/info/files/european-green-deal-communication_en.pdf

European Commission. (2019b). *Bioeconomy policy*. https://ec.europa.eu/research/bioeconomy/index.cfm?pg=policy&lib=strategy

European Commission. (2020). *Horizon 2020 Programme: Funding & tender opportunities*. http://ec.europa.eu/research/participants/portal/desktop/en/home.html

H2OCEAN. (2012). *FP7 European Commission project*. Online platform. https://cordis.europa.eu/project/id/288145. Accessed 18 Apr 2018.

IMO MARPOL Annex VI. (2010). *Prevention of air pollution from ships*. http://www.imo.org/en/OurWork/Environment/PollutionPrevention/AirPollution/Pages/Air-Pollution.aspx

IPCC. (2018). *Special report: Global warming of 1.5 °C*. https://www.ipcc.ch/sr15/. Accessed 18 Apr 2018.

Koundouri, P. (2017). *The ocean of tomorrow, investment assessment of multi-use offshore platforms: Methodology and applications – Volume 1*. Springer. https://doi.org/10.1007/978-3-319-55772-4. eBook ISBN: 978-3-319-55772-4, Hardcover ISBN: 978-3-319-55770-0.

MSFD. (2008). Marine Strategy Framework Directive. Directive 2008/56/EC of the European Parliament and of the Council of 17 June 2008 establishing a framework for community action in the field of marine environmental policy. *Official Journal of the European Union, L 164, 25.6.2008*, 19–40.

MERMAID. (2015). *Online platform*. http://www.vliz.be/projects/mermaidproject/. Accessed 18 Apr 2018.

MERMAID. (2019). *Public documents*. http://www.vliz.be/projects/mermaidproject/docmanager/public/. Accessed 18 Apr 2018.

Ocean Literacy Network. (2015). *Ocean Literacy Principle #6*. http://oceanliteracy.wp2.coexploration.org/ocean-literacy-framework/principle-6-v2/

RECONNECT. (2020). Interreg European Commission project. Regional cooperation for the transnational ecosystem sustainable development. Online platform. https://reconnect.hcmr.gr

THAL-CHOR. (2015). *Cross border cooperation for Maritime Spatial Planning Development*. Interreg A: Cross-Border Cooperation Programme "Greece-Cyprus 2007–2013. https://www.msp-platform.eu/projects/cross-border-cooperation-maritime-spatial-planning-development

TROPOS. (2012). *FP7 European Commission project*. Online platform. http://www.troposplatform.eu/

Tsani, S., & Koundouri, P. (2018). A methodological note for the development of integrated aquaculture production models. *Frontiers in Marine Science, 4*, 2018. https://doi.org/10.3389/fmars.2017.00406. ISSN: 2296-7745.

UN SDSN. (2020). *United Nations Sustainable Development Solutions Network*. https://www.unsdsn.org/. Accessed on 13 Mar 2020.

UNCTAD. (2019). *Review of Maritime Transport 2019*. https://unctad.org/en/PublicationsLibrary/rmt2019_en.pdf

UNCTAD. (2020). *Climate Change and Maritime Transport*. https://unctad.org/en/Pages/DTL/TTL/Legal/Climate-Change-and-Maritime-Transport.aspx

UNEPMAP. (2020). *Mediterranean 2017 Quality Status Report*. https://www.medqsr.org. Accessed on 5 Apr 2020.

Vassilopoulos, A., & Koundouri, P. (2017). Valuation of marine ecosystems. In *Oxford research encyclopedia of environmental science*. Oxford University Press. https://doi.org/10.1093/acrefore/9780199389414.013.529. e-ISBN: 9780199389414.

WEF. (2016). *The new plastics economy. Rethinking the future of plastics*. http://www3.weforum.org/docs/WEF_The_New_Plastics_Economy.pdf

Chapter 2
Stakeholder Involvement in Technological Design: Lessons Learned from the MERMAID and TROPOS Projects

Marian Stuiver, Sander van den Burg, Wenting Chen, Claire Haggett, David Rudolph, and Phoebe Koundouri

Abstract Shared multi-use of ocean space is associated with overcoming several complex technical, regulatory, financial, environmental and socio-economic problems. In achieving this goal, several stakeholders of relevance need to participate in the design and implementation of multi-use platforms. This chapter (This work has received funding from the European Union's 7th Framework Program under grant agreement N° 288710 and N° 288192) reviews and discusses the participatory approaches employed in the MERMAID and TROPOS projects. The discussion draws on the methods employed in each case, the objectives and obstacles encountered resulting in useful conclusions for participatory design.

Keywords MERMAID · TROPOS · Multi-use platforms · Stakeholder engagement

---

M. Stuiver
Wageningen Environmental Research at Wageningen, Wageningen, The Netherlands

S. van den Burg
Wageningen University & Research, Wageningen, The Netherlands

W. Chen
Norwegian Institute for Water Research (NIVA), Oslo, Norway

C. Haggett
School of Social and Political Science, University of Edinburgh, Edinburgh, UK

D. Rudolph
Department of Wind Energy, Technical University of Denmark, Kongens Lyngby, Denmark

P. Koundouri (✉)
School of Economics and ReSEES Laboratory, Athens University of Economics and Business, Athens, Greece

United Nations Sustainable Development Solutions Network Europe, Paris, France

EIT Climate–KIC Hub Greece, ATHENA Research and Innovation Center, Athens, Greece

ICRE8: International Centre for Research on the Environment and the Economy, Athens, Greece
e-mail: pkoundouri@aueb.gr

P. Koundouri (ed.), *The Ocean of Tomorrow*, Environment & Policy 57,
https://doi.org/10.1007/978-3-030-56847-4_2

2.1 Introduction

The European Commission has launched the Blue Growth Strategy, being the EU's long-term strategy to support sustainable growth in the marine and maritime sectors. Much is expected from growth of the maritime sectors aquaculture, renewable energy, blue biotechnology, seabed mining and tourism, contributing to the economic growth and creation of new jobs (Stuiver et al. 2012). If this growth takes place, the distribution of marine space will become a delicate issue in European seas. These activities require space. Although the maritime areas are large, these activities often eye for the same sites, close to the coastline with good access to facilities.

This results in competition for space, already visible in (among others) the North Sea area. Solving this problem requires maritime spatial planning and the development of creative and innovative contributions which at the same time can ensure environmental sustainability. One such creative and innovative possibility is multi-use platforms, when two or more sectors join activities in same area, or even using the same infrastructures. While multi-use platforms are accepted as a valuable contribution to Blue Growth, the design and employment of multi-use platforms is complicated. This is due to the fact that they involve multiple knowledge questions, challenges for governance and issues of sustainability and that they require cross-sectoral cooperation.

To overcome these complex problems – where technical, regulatory, financial, environmental and socio-economic aspects are intertwined – stakeholders need to be participating in the design and implementation of multi-use platforms. This is to make sure that the designed solutions are not developed top-down by science or policymakers but that they built on the knowledge and experience of various stakeholders to come to feasible and sustainable solutions (Wenting et al. 2014, 2015). Comparable pleas for stakeholder participation are made in the European directives on marine spatial planning (Aitken et al. 2016). A participatory approach replaces linear models of knowledge generation which focus on scientists and engineers in the design of new technologies. Inclusion of stakeholders from different backgrounds can enable handling complex problems with multiple interests and objectives (Stuiver et al. 2012; Pomeroy and Douvere 2008).

Participatory design values the perspective, knowledge, skills and involvement of different categories of end users and other stakeholders (Wilkinson and De Angeli 2014; Murgue et al. 2015; Simonsen and Robertson 2012).[1] Participatory design also embraces the two principles of non-linear knowledge generation and social learning. The first principle acknowledges that knowledge is developed in a complex, interactive process of co-production with a range of stakeholders involved. The second principle states that all one can do in complex and uncertain search processes for sustainable designs with no ready-made solutions at hand is to experiment and learn

[1]Other participatory design studies have focused on a broad range of subjects such as health information systems (Wesselink et al. 2011) and medical devices such as wheelchairs (Berghöfer et al. 2008), but also on agricultural landscapes (Pilemalm and Timpka 2008).

from these experiments in a social environment through interaction with other actors and through learning from each other's behaviour.

This chapter compares the participatory approaches employed at the MERMAID and the TROPOS projects that studied the development of multi-use platforms. We describe the approach taken, and we relate them to the objectives of participation with the aim to assess if participation was valuable. The chapter draws useful conclusions of participatory design, based on the findings of the two projects with the aim to inform future design processes. The remainder of the chapter develops as follows: Sections 2.2 and 2.3 discuss the importance and the approach to stakeholder involvement in the TRPOPOS and MERMAID projects, respectively. Section 2.4 provides a comparative assessment, and the last section concludes with useful recommendations for future applications.

2.2 Stakeholder Involvement in TROPOS

2.2.1 *The Importance to Involve Stakeholders*

The TROPOS project aimed to develop a floating modular multi-use platform system for use in deep waters with an initial geographic focus on the locations off Crete (Greece), Canary Islands and Taiwan. The TROPOS multi-use platform system wanted to integrate a range of functions including the transport and aquaculture sectors, leisure activities and renewable energy production. The so-called Leisure Island concept was planned for Gran Canaria (Canary Islands), with a combination of solar energy production with offshore leisure activities. For Taiwan a so-called Green & Blue platform concept was planned as a future concept. In this scenario 'ocean thermal energy conversion' (OTEC) was integrated with offshore aquaculture. For Crete, the Green & Blue platform concept was also planned. Different from that in Taiwan, offshore wind energy is combined with offshore aquaculture. A fourth scenario, the Sustainable Service Hub Concept, was located on the Dogger Bank in the North Sea and has been recently included in the scope of TROPOS (Wenting et al. 2014).

Stakeholder involvement in the TROPOS project focused on social acceptance. Social acceptance is recognised as a crucial issue in shaping a widespread and successful implementation of novel technologies and infrastructures. Research on other developments (such as offshore wind farms) has demonstrated that local opposition can delay and prevent developments; and, conversely, where the public and key stakeholders are engaged and consulted, this results in a better project, which can demonstrate an understanding and awareness of the local geographical, social, economic and political context (Aitken et al. 2016).

In the TROPOS project, social acceptance was regarded as a dynamic process driven by various and alterable values rather than being based on static 'facts', which also refers to issues of power, procedural justice, governance, communication and engagement (Haggett 2011; Hall et al. 2013). Since the TROPOS platform was meant to be relatively close to the shore and since the concept included offshore solar, wind farm and OTEC, valuable lessons on social acceptance can be drawn from the rich literature on acceptance of offshore renewables. In the TROPOS study, the concepts of Leisure Island concept off Gran Canaria and the Green & Blue concept off Liuqiu Island, Taiwan, were used as examples (Chen et al. 2014).

2.2.2 Stakeholder Groups

Stakeholders were identified as those who will be likely affected by the development of the multifunctional platform and whose interests and concerns should be taken into account in the planning of large-scale offshore platforms. The stakeholder groups comprised members of the following groups: local and regional governments, fishing and shipping communities, local businesspeople, local tourism and leisure industries, environmental organisations and coastal communities.

2.2.3 Activities Performed and Their Outcome

The TROPOS project made use of a combination of qualitative and quantitative approaches when involving the relevant stakeholders. Firstly, the potential socio-economic impacts were identified from the literature. Six main socio-economic impact categories listed in Vanclay (2002) were used in the TROPOS case studies as they are commonly used both in literature and practice. These include health and social well-being impacts, quality of living environment impacts, economic and material well-being impacts, cultural impacts, family and community impacts and institutional legal political equality impacts. Then the impacts were adjusted for each case study during the survey in the local area to fit the local situation better.

Secondly, the strategies for data collection were developed. As the methodology and the choice of methods for data collection should always be guided by the specific situation, the acquisition of empirical data was uniquely created for each case. That is, for Leisure Island concept off Gran Canaria, TROPOS adopted a mixed method approach which comprises a face-to-face survey and a semi-structured questionnaire survey with local people and tourists as well as in-depth interviews with stakeholders who are potentially affected by offshore developments. As far as the Green & Blue concept off Liuqiu Island, Taiwan, is concerned, TROPOS used a face-to-face survey for local people and members of fishing industry. These changes were particularly made to shift the focus away from tourism to put more emphasis on local residents and members of the fishing industry as more significant stakeholders. Thirdly, a pilot survey in each case site was carried out in a smaller scale to check whether the initial questionnaire design needed to be changed to the local condition.

Finally, a final survey was carried out for each case, and these results were analysed. In the survey and semi-structured interviews, the following aspects of social acceptance were investigated: knowledge and awareness of the project, support for the project, perceptions of socio-economic impacts, perceptions on sustainable tourism, biodiversity and contrasting environment concern, likelihood to visit to Leisure Island and how tourists willingness to visit the area will be affected by the platform.

2.2.4 Results of Stakeholder Involvement

The results from Gran Canaria showed that there are concerns besides a general high acceptance of the Leisure Island among tourists as well as residents. Benefits for the tourist sector which are predicted to result in an increase of income and generation of jobs become confronted with various potential environmental impacts, in particular the disruption of marine species and habitats.

The results from Liuqiu Island, Taiwan, also demonstrated a generally high acceptance of the Green & Blue platform among residents and tourists, although most participants had been unaware of the project. Despite the general acceptance of such a project, people also raised a number of specific concerns. These concerns are predominantly related to environmental impacts and unclear effects on local fishing and fish processing industries. Other issues that challenged the acceptance of the project include uncertain environmental impacts and adverse effects caused by the operation and construction of the platform. Perceptions of negative impacts were balanced against potential benefits for tourism, which is a crucial economic driver for Liuqiu Island.

This calls for a further examination of local concerns and how people make sense of the interplay between the use of the platform and existing fisheries in order to determine how both may inform, complement and exclude each other.

2.3 Stakeholders' Involvement in MERMAID

2.3.1 The Importance to Involve Stakeholders

The MERMAID project had the aim to develop concepts for the next generation of offshore activities for multi-use of ocean space. The project did not envisage to actually implement these activities, but it examined new design concepts for combining offshore activities like energy extraction, aquaculture and platform-related transport at various areas in the ocean. In order to achieve this, the MERMAID project put the integration of technical, economic, ecological, spatial and social aspects at the heart of the design of MUPs in two ways: first by analysing and integrating all these aspects in the design and second by involving stakeholders in the entire design process. For the latter, a participatory design process was developed (Rasenberg et al. 2013). The objective of the participatory design process was to work together with the users and other relevant stakeholders throughout the design and development process. For this purpose, a participation process was executed throughout the MERMAID project that focuses on a cyclical, iterative and participatory process of scoping, envisioning and learning through which a shared interpretation of MUPs was developed and applied in an integrated manner.

2.3.2 Stakeholder Groups

A group of representatives of all major types of stakeholders were invited for the interviews and round table sessions, where six stakeholder categories were identified: governing bodies/policymakers such as regional, national and European officers; end users of the multi-use platforms, e.g. energy companies and aquaculture entrepreneurs; suppliers of the multi-use platforms such as cable companies and construction businesses; representatives of other offshore activities such as fisheries, shipping and mining sectors; civil society, including (environmental) NGOs and local citizens; and universities and research institutes.

2.3.3 Activities Performed and Their Outcome

The participatory design was developed to involve various stakeholders, with different backgrounds, in the design of a multi-use platform for a specified site. The participatory process was thus performed four times. The four case studies were chosen during the first phase of the MERMAID project. They differed not only in terms of the geophysical characterisation, the degree to which multi-use was discussed before the project also differed. The four sites were the Baltic Sea, a typical estuarine area with fresh water from rivers and salt water, where multi-use was discussed before; the transboundary area of the North Sea and Wadden Sea, a typical active morphology site, where multi-use was discussed before; the Atlantic Ocean, a typical exposed deepwater site, where multi-use was not discussed before; and the Mediterranean Sea, a typical sheltered deepwater site, where multi-use was not discussed before.

Figure 2.1 gives an overview of the participatory design process which was applied in these four case studies in the MERMAID project. The design process of MUPs in the four cases was organised in three steps as follows: first prepare the designs by identifying the views and needs of all stakeholders: this was done through interviews (Rasenberg et al. 2013); then give inputs for the design of the multi-use platform organising a round table session involving all stakeholders (Röckmann et al. 2015); and finally evaluate the design by organising a round table session with all stakeholders (Röckmann et al. 2015).

The work performed in the participatory process did not aim to make a final design, but to organise the input of the stakeholders that can be used to make the final

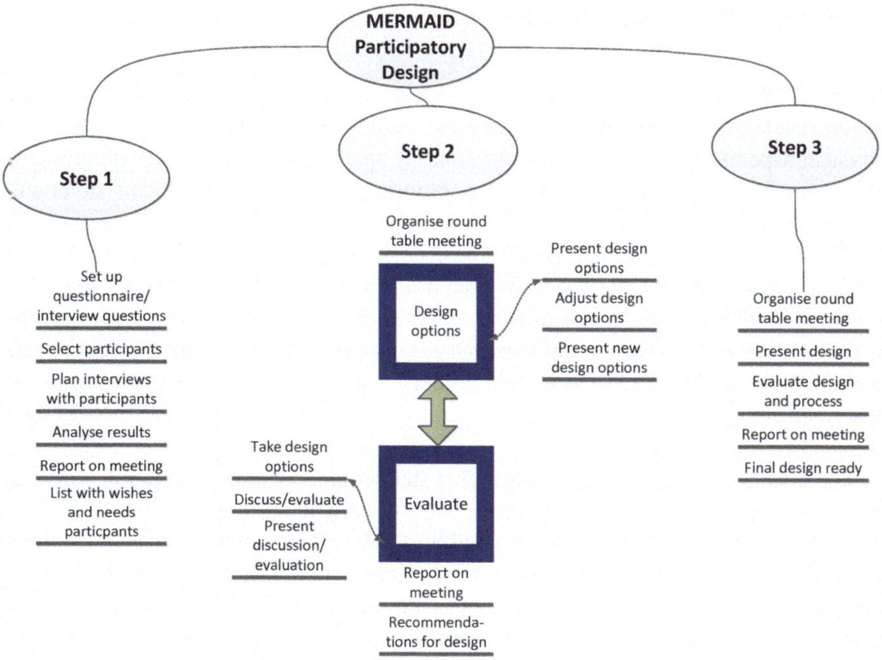

Fig. 2.1 Overview of the MERMAID participatory design process

design. The final design was the responsibility of the site managers (each of the sites has a site manager) for the different case studies. The site managers also played a crucial role in organising the three steps of the participatory design.

The first step took place in 2012, and the results of this step are reported in Rasenberg et al. (2013). In the first step, interviews were held with representatives of a wide range of stakeholders. First step focused on identifying different views on ecological, economic and social objectives of multi-use platforms, challenges and technical, social-economic and ecological constraints faced. Equipped with a resulting wish list from this step, designers started working on developing the first design options. These design options were discussed later in the second step in an interactive round table session involving all relevant stakeholders.

The second step constituted of an iterative cycle where draft design options were presented and developed. The information provided valuable input to the designers that were responsible for the final design. Based on the discussions in the round table sessions on these design options with regard to ecological, economic, social, technical and governance aspects, the design options were translated into a final design concept.

Last step constituted of a round table session where the final design concept was evaluated with the participating stakeholders. This ultimately led to a design concept which was thoroughly analysed, technically feasible and preferably supported by all the stakeholders represented at the round table.

2.3.4 Results of Stakeholder Involvement

The MERMAID project focused on four case study sites representative for European waters, each with local challenges. The case studies differed not only with respect to physical aspects but also with respect to marine spatial planning, current planning of offshore wind development, aquaculture activities and governance. The involvement of stakeholders also differs at some sites several stakeholders were part of the MERMAID project (e.g. the Baltic site), whereas in other cases, there were no local projects partners that also were stakeholders (e.g. the Mediterranean site).

The involvement of the stakeholders resulted in four different designs for multi-use platforms at sea. The outcome of the design process, which stakeholders were present and how the participating stakeholders' concerns were taken into account, is described in greater detail in Rasenberg et al. (2013, 2014) and Röckmann et al. (2015).

The results of the North Sea showed that stakeholders who had a relation with this area – for current and potential future uses – were identified, and the following stakeholders participated: offshore wind farm developers, seaweed and shellfish aquaculture fisheries, regulators and policymakers, offshore construction companies, companies interested in the end product of the design and NGOs. In collaboration with the stakeholders, offshore wind parks combined with seaweed and mussel aquaculture were identified as the most promising conceptual multi-use design. The final design did not fully integrate the aquaculture structures inside the wind farm, but instead in the areas just outside of and in between the wind farms.

Many of the participating stakeholders could see benefits in participating in a multi-use platform, e.g. with regard to more efficient use of space and functional synergies. The idea was not new to stakeholders, and their discussion focused on optimisation with regard to sharing infrastructures to reduce O&M costs and create win-win solutions. To increase employment and support the fisheries sector, their vessels, possibly redesigned as part of an infrastructure, was seen as an important aspect to consider. The biggest challenge for the North Sea site was to find solutions that could be profitable for all stakeholders, including risks and extra insurance costs. The wind energy developers were very clear regarding the conditions for design: multi-use should not cause any hindrances for wind turbines or obstacles for operation and maintenance activities. Furthermore, it was stated that in order to find investors, the license procedures need to be aligned for multi-use, i.e. faster than today, and uncertainties need to be minimised (van den Burg et al. 2016).

The results of the Mediterranean showed that stakeholders included the energy sector, aquaculture, policymakers and environmental authorities. The final design for the Mediterranean site included grid-connected wind turbines combined with fish farming. In the layout, the fish farm is placed in the space among the wind turbines, leading to a total occupied space of 0.64 km^2 (Röckmann et al. 2015).

A number of stakeholders initially opposed to the idea of including aquaculture farms in the multi-use platforms, because they were afraid of competition with the already existing coastal aquaculture. Despite this fear of competition, the design team decided not to limit the design by this argument, as this essentially was a plea for keeping a monopoly of the coastal aquaculture. Aquaculture was considered and included in the proposed design, since it is an activity that can be combined with the other uses. An additional supporting argument for including aquaculture in the proposed design was the existence and vicinity of a market for aquaculture products nearby.

In the Atlantic case, the invited stakeholders included offshore energy sector, aquaculture, suppliers to the offshore industry as well as NGOs and scientists. The final design included a combination of floating offshore wind turbines and wave energy generators. Stakeholders argued for the importance to select a site where conflicts with other interests are minimal, e.g. a platform should be sufficiently far away from the coast, and not cause negative impacts on the local fishing community. Multi-use platforms were considered to be able to provide revenues to both the local fishing community and local businesses. The importance of including marine renewable energy technology in the design, and the benefits of this sector in the area of Cantabria, was agreed upon.

During the round table meeting, the aquaculture sector showed interest in the development of a platform. However, after discussions with all the stakeholders, aquaculture was deemed very difficult technically. The discussion identified the need for cooperation between stakeholders for accurate designs. Some respondents provided examples to illustrate its importance: technically well-designed projects can still run into problems. Economic issues were also identified as a way to integrate MUPS in the local society: MUPS development may lead to the creation of new jobs in the area.

In the Baltic case, stakeholders included potential entrepreneurs to participate in the development of a multi-use platform, as well as governing bodies and the shipping authorities. Also, NGOs representing societal values and scientists participated. The eventual design combined wind turbines and off-shore aquaculture by floating fish cages with trout/salmon production. This combination was interesting given the large-scale development of offshore wind – with subsequent spatial claims and the critical attitude towards nearshore aquaculture.

The stakeholders lifted forward social aspects with regard to the visibility of wind turbines. However, the design and location were such that, depending on the weather conditions, the wind turbines will seldom be visible. The entire wind park area should ideally be designated a cable protection area, and possibly shipping lines which today pass the area need to be altered. Stakeholders discussed technical aspects for design such as maintenance and monitoring, anchoring and transport and associated risks. A technical risk assessment of the multi-use platforms appeared to be important, and guidelines and rules to minimise risks must be developed to ensure the safety of people, vessels, cages and wind turbines. The stakeholders pointed out that there should be no negative effects on ecological conditions and that the artificial reefs on the wind turbines foundations should be protected as they have positive ecological effects.

2.4 Comparison of the Two Approaches

We proceed with the comparison of the objectives of participation of the two different projects looking in particular at whether the projects have succeeded in this endeavour. Table 2.1 provides a summary of the approaches.

TROPOS project engaged a range of stakeholders in the research strategy. This allowed the scientists to understand the different perspectives of those who had a particular interest in the area, as well as those who chose to visit, and those who lived nearby, using a multi-method strategy which aimed to understand in depth the views of stakeholders and a breadth of opinion from residents and tourists.

As with all research projects of this nature, finding the right respondents is a key challenge. There is always a risk of 'selection bias' that those who participate are those who are willing to do so and that those who refuse may hold views which are different and which are therefore excluded from the analysis. TROPOS tried to minimise this bias where at all possible, by stressing the very limited involvement (a short amount of time to participate, convenient to the interviewee), and making the research and the project seem as interesting and engaging as possible, to encourage residents and tourists to participate.

Researching a hypothetical development is always challenging also; it is much more straightforward for researchers and participants to discuss something that already exists. Therefore, the respondents were asked about their expectations, and his became part of the analysis. It was also important to give as much information as possible regarding the platforms, without seeming to try and push them in one direction or another in their views. Future research could benefit from projects which were more widely known about or further along in their stage of development, which would make discussing them and their impacts more straightforward.

In MERMAID and its four case studies, the involvement of stakeholders has made a valuable contribution to the design of the platforms in the four study sites. In all cases, there has been an exchange of knowledge, interest and at least a start in bringing different sectors together. Dominant concerns – such as the weather and wave condition in the Atlantic site – were brought to the table and could be addressed in the designs.

A weak point of the approach taken is that the current status quo – whether that concerns market organisation and technologies available or fits with policy objectives – proved to very influential in the design process; e.g. the Mediterranean stakeholders were reluctant to include aquaculture in the design because aquaculture is nowadays dominated by coastal aquaculture entrepreneurs. It proved to be difficult to apply 'blue sky thinking' to these cases.

The most valuable lesson is that the role of stakeholders will differ per case and that consequently the selected approach should be tailored to the situation. In the Baltic site, a predefined group of MERMAID participants sought to discuss the feasibility of realising a multi-use platform at a specific location. This is different from the Atlantic and Mediterranean sites, where the idea was unknown beforehand, and the process was aimed at bringing together stakeholders to explore the potential

Table 2.1 Comparison of the participatory design in TROPOS and MERMAID

	TROPOS	MERMAID
Objective	Develop social acceptance of multi-use offshore platform in the local area	Design socially accepted concepts for multi-use platforms
Stakeholders involved	Local and regional government	Governing bodies/policymakers End users of the MUP
	Fishing and shipping communities	Suppliers
	Local businesspeople	Representatives of other offshore activities
	Local tourism and leisure industries	Environmental NGOs, local citizens
	Environmental organisations	Universities and research institutes
	Coastal communities.	
Methodology	A combination of qualitative and quantitative method, i.e.:	Interactive design
	Face-to-face survey	Interviews
	A semi-structured questionnaire survey	Round tables
	In-depth interviews	Collect knowledge and information from the stakeholders
		Eventual design was the responsibility of the project team
Results	Social acceptance of the platforms in both case studies	4 designs for multi-use platforms based on stakeholder input
	People show concern to the environmental impacts of the platform	Acceptance of design among the stakeholders
	Stakeholders are positive to the economic benefits generated by the platform	Understanding of the complexity of multi-use platform development
		People show concern to the environmental impacts of the platform
Strong points	The fusion of the qualitative and quantitative epistemologies complemented each other	Design approach useable in context where MUPS were not known beforehand
	Research process gained social acceptance	Broad involvement of stakeholders
		Learning processes on multi-use were given a boost in the areas
Weak points	Need to be mindful to avoid selection bias when finding respondents	Generic approach should be tailored to different situations in different basins
	Need to maintain a balance of information without biasing respondents one way or the other when presenting information to them	Difficult to make steps towards implementation
		Barriers to 'blue sky thinking' are present in the real-life situation
Value of participation	People perceive and ascribe meaning. This is decisive for the successful development and implementation of these infrastructures	Incorporation of local knowledge is important in the designs, and learning is encouraged among all the stakeholders
	Developers and decision-makers need to consider the legitimacy of new design and learn about their potential concerns/challenges	It is a start of bringing sectors together that have to implement the designs in the future

of multi-use platforms at these locations. The North Sea site was in between; even though the site was predetermined, the stakeholders embarked in a process to better understand each other's needs in developing multi-use platforms at the specified location.

2.5 Concluding Remarks

MERMAID and TROPOS point to the importance of a thorough consideration of stakeholders' concerns in design processes to take economic actors, government, local citizens and others on board, to legitimise the project and to integrate the platform more effectively in the local context. One has to consider that, therefore, stakeholder involvement requires a considerable investment from all the participants in the project.

When setting up participation, different choices need to be made in different stages of stakeholder involvement. If the process is aimed at closing the deal, as was visible in the Baltic case study of MERMAID, other actions are needed, then when the aim is to make stakeholders aware of the potential of multi-use, such as in the Atlantic case study of MERMAID, or in the TROPOS projects. It is important to investigate what project phase applies to the proposed site, e.g. identifying a business case, exploring options to 'add' functions to a planned development or investigating the idea of multi-use platforms from scratch.

It is very important to know the situation and conditions of the site under consideration – e.g. what technologies are at all possible. Therefore, members of the team need to invest a lot of time in understanding the locality of the case studies. In the future, other projects should provide the necessary resources for creating this understanding of the locality as it is crucial for identifying the relevant stakeholders, their roles, objectives and resources.

When eliciting stakeholders' view, selection bias should be avoided during both the preparation and interview stages. It is recommended to involve the relevant set of stakeholders for specific decisions. In early exploratory project phases, take stock of differing views of the stakeholders. In a technical scoping phase, it makes sense to only involve a small group of relevant experts. In later project phases, stakeholders should be asked to pronounce themselves on few and reasonably well-defined design options that are possible for the specific offshore multi-use platform.

Finally, shared knowledge and experience can contribute to more efficient and sustainable designs of offshore multi-use platforms. Taking into account stakeholders' perspective and knowledge can help overcome obstacles, and lead one to adjust the design process if necessary. Not engaging in dialogue or not considering stakeholders' point of view leads to risk of inefficient processes, the need to repeat procedures or even implement suboptimal solutions.

References

Aitken, M., Haggett, C., & Rudolph, D. (2016). Practices and rationales of community engagement with wind farms: Awareness raising, consultation, empowerment. *Planning Theory and Practice, 17*(4), 557–576.

Berghöfer, A., Wittmer, H., & Rauschmayer, F. (2008). Stakeholder participation in ecosystem-based approaches to fisheries management: A synthesis from European research projects. *Marine Policy, 32*, 243–253. https://doi.org/10.1016/j.marpol.2007.09.014. (07-12-2015).

Chen W., Ruldoph, D., Haggett, C., Seifert-Dähnn, I., Golmen, L., Daria, J., Quevedo, E., Grito, J. H., Ying, F., Lu, S. Y., & Mintenbeck, K. (2014). *D6.4 A framework for describing the social impacts with concrete examples that apply for the Canary Island, EU FP7 TROPOS project.*

Haggett, C. (2011). Planning and persuasion: Public engagement in renewable energy decision-making. In P. Devine-Wright (Ed.), *Renewable energy and the public: From NIMBY to participation* (pp. 15–27). Earthscan: London.

Hall, N., Ashworth, P., & Devine-Wright, P. (2013). Social acceptance of wind farms: Analysis of four common themes across Australian case studies. *Energy Policy, 58*, 200–208.

Murgue, C., Therond, O., & Leenhardt, D. (2015). Toward integrated water and agricultural land management: Participatory design of agricultural landscapes. *Land Use Policy, 45*, 52–63.

Pilemalm, S., & Timpka, T. (2008). Third generation participatory design in health informatics—Making user participation applicable to large-scale information system projects. *Journal of Biomedical Informatics, 41*(2), 327–339.

Pomeroy, R., & Douvere, F. (2008). The engagement of stakeholders in the marine spatial planning process. *Marine Policy, 32*, 816–822. https://doi.org/10.1016/j.marpol.2008.03.017. (07-12-2015).

Rasenberg, M., Stuiver, M., Van den Burg, S., Veenstra, F., Norrman, J., & Söderqvist, T. (2013). *Stakeholder views; deliverable D2.2, MERMAID project.*

Rasenberg, M., Stuiver, M., Van den Burg, S., Norrman, J., & Söderqvist, T. (2014). *Stakeholder views 2; deliverable D2.3, MERMAID project.*

Röckmann, C., Stuiver, M., van den Burg, S., Zanuttigh, B., Zagonari, F., Airoldi, L., Angelelli, E., Suffredini, R., Franceschi, G., Bellotti, G., Schouten, J.J., Söderqvist, T., Garção, R., Guanche Garcia, R., Sarmiento Martínez, J., Svenstrup Petersen, O., & Aarup Ahrensberg, N. (2015). Platform solutions; deliverable 2.4, MERMAID project.

Simonsen, J., & Robertson, T. (2012). *Routledge international handbook of participatory design.* New York: Routledge.

Stuiver, M., Agricola, H. J., Fontein, R. J., Gerritsen, A. L., Kersten, P. H., & Kselik, R. A. L. (2012). *Multifunctionele Platforms op Zee, het concept, de wet en regelgeving en de lessen voor de toekomst.* Wageningen: Alterra, (Alterra-rapport 2364).

van den Burg, S. W. K., Stuiver, M., Norman, J., Garcao, R., & Rockmann, C. (2016). Participatory design of multi-use platforms at sea. *Sustainability, 8*(2)., 17 p.

Vanclay, F. (2002). Conceptualising social impacts. *Environmental Impact Assessment Review, 22*, 183–211.

Wenting, C., Ruldoph, D., Haggett, C., Seifert-Dähnn, I., Golmen, L., Daria, J., Quevedo, E., Grito, J. H., Ying, F., Lu, S. Y., & Mintenbeck, K. (2014*). D6.4 A framework for describing the social impacts with concrete examples that apply for the Canary Island, EU FP7 TROPOS project.*

Wenting, C., Koundouri, P., Dávila, O. G.; Souliotis, Y., Haggett, C., Ruldoph, D., Ying, F., Lu, S. Y., Chi, C.; Lin, J.; Li, S., Mintenbeck, K., Gunder, G. L., Quevedo, E., & Grito, J. H., (2015) *D6.6 A framework for describing the social impact with concrete examples that apply for Green and Blue Concept in Taiwan- a joint report between EU FP7 TROPOS and EU FP7 MERMAID, EU FP7 TROPOS project.*

Wesselink, A., Paavola, J., Fritsch, O., & Renn, O. (2011). Rationales for public participation in environmental policy and governance: Practitioners' perspectives. *Environmental Plan A, 43*, 2688–2704. (07-12-2015).

Wilkinson, C. R., & De Angeli, A. (2014). Applying user centred and participatory design approaches to commercial product development. *Design Studies, 35*(6), 614–631.

Chapter 3
Comparative Financial Analysis of Marine Multipurpose Platforms Projects: MERMAID and TROPOS Projects

Saúl Torres-Ortega, Pedro Díaz-Simal, Fernando Del-Jesus, Raúl Guanche, and Phoebe Koundouri

Abstract The Oceans of Tomorrow (FP7-OCEAN) initiative aimed to foster multidisciplinary approaches between different economic and scientific sectors and disciplines, maintaining a common focus on marine and maritime challenges. Under this umbrella, three projects were developed focusing on the development of multi-use offshore platforms: H2Ocean, TROPOS and MERMAID. The development of all these three projects included the design of different concepts in terms of study cases and application of the main findings of the projects. The financial aspect of these design concepts was carried out in all projects, but the assumptions made to perform this analysis were quite different among the projects. Although none of the Oceans of Tomorrow projects had the objective to produce a viable proposal from the economic point of view, it is necessary to analyse the possibilities of the concepts developed. A common methodology and parameters need to be used in order to achieve a constancy and homogeneity that allows comparing the results. This chapter (This work has received funding from the European Union's 7th Framework Program under grant agreement N° 288710 and N° 288192) defines a common financial framework that permits to obtain comparable results of the financial performance of the different design concepts proposed in the Oceans of Tomorrow projects.

Keywords H2Ocean · TROPOS · MERMAID · Financial framework · Multi-use offshore platform

---

S. Torres-Ortega · P. Díaz-Simal · F. Del-Jesus · R. Guanche
Environmental Hydraulics Institute, Universidad de Cantabria, Santander, Spain

P. Koundouri (✉)
School of Economics and ReSEES Laboratory, Athens University of Economics and Business, Athens, Greece

United Nations Sustainable Development Solutions Network Europe, Paris, France

EIT Climate–KIC Hub Greece, ATHENA Research and Innovation Center, Athens, Greece

ICRE8: International Centre for Research on the Environment and the Economy, Athens, Greece
e-mail: pkoundouri@aueb.gr

P. Koundouri (ed.), *The Ocean of Tomorrow*, Environment & Policy 57,
https://doi.org/10.1007/978-3-030-56847-4_3

3.1 Introduction

This chapter presents the results of a comparative financial analysis performed to the three Oceans of Tomorrow projects. Each one of the projects has generated different deliverables where the economic data are reviewed. However, the direct comparative of these outputs presented in project's deliverables shows diverse weakness. Firstly, different financial indicators were obtained. Secondly, different initial conditions were used. Thirdly, different parameters were used in the analysis. These three problems combined result in the final output not being comparable as it is.

To address this issue and to allow for a clear comparative analysis of the results from the projects, some further actions need to be taken. The methodology, presented in Sect. 3.2.2, includes two main steps. At a first stage, a normalisation of project's data is performed. This normalisation transforms the main financial outflows and inflows in a relative value using the production size of each project as a normalising parameter. These values are then compared to main figures obtained in the scientific literature and other economic information. This step should help to assess if the cost and revenue structure of each project are included in common range of variation. These results are presented in Sect. 3.2.3.

The second stage involves the calculation of a profitability indicator, in this case the net present value. This calculation is performed with standard and homogeneous parameters (such as horizon time, discount rate and construction years) for all different projects. The use of these common boundary conditions makes possible to compare the results of the projects under the same framework. This analysis is detailed in Sect. 3.4. Beyond the data used in the financial analysis, the possibility exists for variability on the values used in the analysis. To assess the impact of the possible variability of the inputs on the financial model, a sensitivity and a risk analysis are also performed. These results are presented in Sects. 3.5 and 3.6, respectively. Finally, some concluding remarks and caveats about the results obtained are included in Sect. 3.7.

3.2 Methodology

The following methodology has the objective to assess and compare the financial viability of multi-use of space and multi-use platform concepts developed on the Oceans of Tomorrow projects studied in H2Ocean, TROPOS and MERMAID.

The methodology to apply is divided into five steps:

1. Review available information from projects' plans.
2. Homogenise available information.
3. Perform financial assessment.
4. Perform sensitivity analysis.
5. Perform risk assessment.

The main outputs of this process are obtained from steps 3, 4 and 5.

- From step 3, the financial net present value (FNPV) is obtained as an indicator of the project profitability. Also, the financial rate of return (FRR) is calculated as a complementary indicator.
- From step 4, the list of the critical parameters of each project is obtained. These parameters are defined as those whose variations, be they positive or negative, have the largest impact on the project's financial performance.
- From step 5, a probabilistic distribution of the FNPV is obtained by assigning a probability distribution to each of the critical variables of the sensitivity analysis.

For the calculation of FNPV and even the sensitivity analysis, Excel is the broadest and commonly tool used. Nevertheless, for the risk assessment, the employment of Monte Carlo simulations is recurrent. For this process, R language programming has been used. The use of an open language makes it easy to completely reproduce an analysis (starting from the same data, we can achieve the same results), so that different researchers will obtain the same results and so that analyses can be audited (Hopper 2016).

The five steps of the methodology are described in detail next.

3.2.1 *Review Available Information from Projects' Plans*

In this first step, all available information from Ocean of Tomorrow projects is gathered so as to obtain the financial and economic data necessary for the assessment. Public deliverables and scientific publications are reviewed looking for this data. When the financial parameters of interest could not be found, or there is evidence that some necessary information is available from not public deliverables, contact with the person responsible of the project has been made asking for their cooperation and collaboration.

3.2.2 *Homogenise Available Information*

This step is necessary due to the different level of detail of information in the different projects. The available information is classified according to the maximum possible detail, and it is grouped in a manner that allows performing the later assessment between projects in time. The use of homogeneous classification of costs and incomes allows to use a common methodology and to compare the obtained results for each project.

3.2.3 Financial Assessment

According to the EC "Guide to Cost-Benefit Analysis (CBA) of Investment Projects" (Sartori et al. 2015), financial assessment has different objectives. These are identified as follows:

- Assess the project profitability.
- Assess the project profitability for the project owner and some key stakeholders.
- Verify the project financial sustainability.
- Outline the cash flows which underpin the calculation of the socio-economic costs and benefits.

To achieve these aims, there are different categories of cash inflows and outflows that must be considered. The most common are described in Table 3.1.

The analysis that the cited guide proposes is based in the use of the discounted cash flow method. This methodology assumes several points:

- Only cash inflows and outflows are considered in the analysis, i.e. depreciation, reserves, price and technical contingencies, and other accounting items which do not correspond to actual flows are disregarded.
- Financial analysis should, as a general rule, be carried out from the point of view of the infrastructure owner. If, in the provision of a general interest service, the

Table 3.1 Inflows and outflows proposed to be included in the financial analysis (Sartori et al. 2015)

Inflows/Outflows	Examples
Investment costs	Start-up and technical costs
	Land
	Buildings
	Equipment
	Machinery
	Replacement costs
	Residual value
Operating costs	Personnel
	Energy
	General expenditure
	Intermediate services
	Raw materials
Other outflows	Loan repayments
	Interests
	Taxes
Inflows	Revenues
	Operating subsidies
Sources of financing	Union assistance
	Public contribution
	Private equity
	Private loan

owner and the operator are not the same entity, a consolidated financial analysis, which excludes the cash flows between the owner and the operator, should be carried out to assess the actual profitability of the investment, independent of the internal payments.

- An appropriate financial discount rate (FDR) is adopted in order to calculate the present value of the future cash flows. The financial discount rate reflects the opportunity cost of capital from the public point of view.
- Project cash flow forecasts should cover a period appropriate to the project's economically useful life and its likely long-term impacts. The number of years for which forecasts are provided should correspond to the project's time horizon (or reference period). The choice of time horizon affects the appraisal results.
- The financial analysis should usually be carried out in constant (real) prices, i.e. with prices fixed at a base year.
- The analysis should be carried out net of VAT.

Project profitability and financial viability are measured by two indicators:

- The financial net present value (FNPV) on investment is defined as the sum that results when the expected investment and operating costs of the project (discounted) are deducted from the discounted value of the expected revenues:

$$FNPV = \sum_{t=0}^{n} \alpha_t S_t = \frac{S_0}{(1+i)^0} + \frac{S_1}{(1+i)^1} + \frac{S_2}{(1+i)^2} + \ldots + \frac{S_n}{(1+i)^n}$$

where S_t is the balance of cash flow at time t, α_t is the financial discount factor chosen for discounting at time t and i is the financial discount rate.

- The financial rate of return on investment is defined as the discount rate that produces a zero FNPV, i.e. FRR is given by the solution of the following equation:

$$0 = \sum_{t=0}^{n} \frac{S_t}{(1+FRR)^t}$$

The FNPV(C) is expressed in money terms (Euro) and must be related to the scale of the project. The FRR(C) is a pure number and is scale-invariant. Mainly, the examiner uses the FRR(C) in order to judge the future performance of the investment in comparison to other projects, or to a benchmark required rate of return.

For the purposes of this analysis, several simplifications and assumptions for the generic methodology are carried out. Homogeneous parameters, as to be able to distinguish and compare the different projects, are proposed. The same is done with the simplification of the flows to be considered.

- Only the flows described in Table 3.2 are used for the analysis. These include CAPEX, OPEX, DECEX, loan flows and revenues.

Table 3.2 Financial flows considered in the analysis

Inflows/outflows	Concept
Investment costs	CAPEX
Operating costs	OPEX
Decommission costs	DECEX
Inflows	Revenues

- An 8.9% is adopted as uniform discount rate for all projects.
- A 4-year construction period is considered, plus 20 years for operation and 1 year for decommission.

Another relevant parameter calculated in addition to NPV and IRR is the levelised cost of production (LCoP). This can be seen as a financial assessment of the average total cost to build and operate an investment over its lifetime distributed over total output produced during the lifetime of the investment considering the discount effect of each unit contribution. The LCoP can also be understood as the minimum cost at which an output must be sold in order to break even over the lifetime of the project.

$$\mathrm{LCoP} = \frac{\text{Sum of costs over lifetime}}{\text{Sum of outputs produced over lifetime}} = \frac{\sum_{t=1}^{n} \frac{\mathrm{CAPEX}_t + \mathrm{OPEX}_t + \mathrm{DECEX}_t}{(1+i)^t}}{\sum_{t=1}^{n} \frac{\mathrm{Output}_t}{(1+i)^t}}$$

3.3 Sensitivity Analysis

Sensitivity analysis enables the identification of the "critical" variables of the project. Such variables are those whose variations, be they positive or negative, have the largest impact on the project's financial and/or economic performance. The analysis is carried out by varying one variable at a time and determining the effect of that change on the NPV. As a guiding criterion, the recommendation is to consider "critical" those variables for which a variation of ±1% of the value adopted in the base case gives rise to a variation of more than 1% in the value of the NPV (Sartori et al. 2015). The tested variables should be deterministically independent and as disaggregated as possible. Correlated variables would give rise to distortions in the results and double counting issues.

A particularly relevant component of the sensitivity analysis is the calculation of the switching values. This is the value that the analysed variable would have to take in order for the NPV of the project to become zero, or, more generally, for the outcome of the project to fall below the minimum level of acceptability. The use of switching values in sensitivity analysis allows making some judgements on the risk of the project and the opportunity of undertaking risk-preventing actions.

3.4 Risk Assessment

This type of analysis assigns a probability distribution to each of the critical variables of the sensitivity analysis, defined in a precise range of values around the best estimate, used as the base case, in order to recalculate the expected values of financial and economic performance indicators. The probability distribution for each variable may be derived from different sources, such as experimental data, distributions found in the literature for similar cases or consultation with experts. Obviously, if the process of generating the distributions is unreliable, the risk assessment is unreliable as well. However, in its simplest design (e.g. triangular distribution), this step is always feasible and represents an important improvement in the understanding of the project's strengths and weaknesses as compared with the base case.

Having established the probability distributions for the critical variables, it is possible to proceed with the calculation of the probability distribution of the FRR or the NPV of the project. For this purpose, the use of the Monte Carlo method is suggested, which requires a simple computation software. The method consists on the repeated random extraction of a set of values for the critical variables, taken within the respective defined intervals, and then the calculation of the performance indices for the project (FRR or NPV) resulting from each set of the extracted values. By repeating this procedure for a large enough number of extractions, one can obtain a predefined convergence of the calculation as the probability distribution of the IRR or NPV.

The values obtained enable the analyst to infer significant judgements about the level of risk of the project. The result of the Monte Carlo drawings, expressed in terms of the probability distribution or cumulated probability of the IRR or the NPV in the resulting interval of values, provides more comprehensive information about the risk profile of a project. The cumulated probability curve (or a table of values) assesses the project risk, for example, verifying whether the cumulative probability for a given value of NPV or IRR is higher or lower than a reference value that is considered to be critical.

3.5 Normalisation of Project's Flows

In this section, the information about flows in the projects (CAPEX, OPEX, DECEX and revenues) are presented in a "normalised format". This step should help to assess the flows per unit of production. This new presentation of the data is useful in the way that it allows to compare with other similar projects. This type of analysis is performed by sector. The data per sector are presented in the following subsections.

3.5.1 Aquaculture

The normalised values for the aquaculture sector have been obtained from the available data of the projects. Two designs include aquaculture activities in their proposals: TROPOS Aquaculture and Mermaid North Sea (Table 3.3).

Table 3.3 Normalised financial values for aquaculture sector

	Parameter	Normalised value	Reference		Common value range (ref)
TROPOS Aquaculture					
European seabass	Production	152.1 ton/cage year	n.a.	In-project internal data	
	Price	825 €/ton	n.a.	In-project internal data	
Meagre	Production	152.1 ton/cage year	n.a.	In-project internal data	
	Price	1043 €/ton	n.a.	In-project internal data	
Greater Amberjack	Production	146 ton/cage year	n.a.	In-project internal data	
	Price	1365 €/ton	n.a.	In-project internal data	
Total	CAPEX	5094.18 €/ton	n.a.	In-project internal data	1.87–9.41 €/kg (*)[3–6]
	OPEX	7651.71 €/ton year	n.a.	In-project internal data	1.74–7.65 €/kg (*)[3–6]
MERMAID North Sea					
Mussel	Production	48,000 ton/year	n.a.	In-project internal data	
	Price	940 €/ton	n.a.	In-project internal data	
	CAPEX	145.83–229.17 €/ton	[7]	Article data	
	OPEX	177.08–1187.5 €/ton year	[8]	Article data	
Seaweed	Production	80,000 ton/year	[9]	Article data	
	Price	210–600 €/ton	[8, 10]	Article data	
	CAPEX	262.50–5000 €/ton	[8, 10, 11]	Article data	
	OPEX	587.5–850 €/ton year	[8, 10, 11]	Article data	

(*) Data for different species

3.5.2 Energy

Three designs include energy production in their proposals: TROPOS Service hub, Mermaid Atlantic Ocean and Mermaid North Sea. Data are summarised in Table 3.4. In the case of offshore wind OPEX costs, it is important to note that relevant uncertainty that exists about these values, due to the few operating offshore wind parks that exist in the world. In the case of MERMAID Atlantic Ocean design, no common references for values are provided as there is no other known project that combines wind and wave (excepting those research projects, where cost structure is not clearly defined).

Table 3.4 Normalised financial values for energy sector

	Parameter	Normalised value	Reference		Common value range (ref)
TROPOS Service hub					
Wind	Power installed	500 MW	n.a.	In-project internal data	
	Production	1,796,000 MWh/year	n.a.	In-project internal data	
	Price		n.a.	In-project internal data	
	LCoE	1252 €/kWh	n.a.	In-project internal data	100–190 €/MWh [12]
	CAPEX	4,173,000 €/MW	[16]	In-project internal data	1,600,000–4,000,000 €/MW [12–15]
	OPEX	53351.66 €/MW/year	[16]	In-project internal data	0.035 CAPEX [15]
MERMAID Atlantic Ocean					
Wind +Wave	Power installed	616 MW	n.a.	In-project internal data	
	Production	80,000 MWh/year	n.a.	In-project internal data	
	Price (8 first years)	150.0 €/MWh	n.a.	In-project internal data	
	Price (after 8 years)	170.0 €/MWh	n.a.	In-project internal data	

(continued)

Table 3.4 (continued)

	Parameter	Normalised value	Reference		Common value range (ref)
	LCoE	167.00 €/MWh	n.a.	In-project internal data	
	CAPEX (mix)	3,664,683 €/MW	n.a.	In-project internal data	
	OPEX (mix)	46,926 €/MW/year	n.a.	In-project internal data	
MERMAID North Sea					
Wind	Power installed	600 MW	n.a.	In-project internal data	
	Production	2,600,000 MWh/year	n.a.	In-project internal data	
	Price (with subsidies)	170 €/MWh	n.a.	In-project internal data	
	Price (without subsidies)	43 €/MWh	n.a.	In-project internal data	
	LCoE		n.a.	In-project internal data	100–190 €/MWh [12]
	CAPEX	4,666,666 €/MW	[18]	Real project data	1,600,000–4,000,000 €/MW [12–15]
	OPEX	100,000–2,300,000 €/MW/year	[8, 19–21]	Article, report data	0.035 CAPEX [15]

3.5.3 *Leisure*

Only one design includes leisure activities in their proposals: TROPOS Leisure data are summarised in Table 3.5.

3.5.4 *Container Transport*

Only one design includes container transport activity. This is the TROPOS Container terminal (Table 3.6).

Table 3.5 Normalised financial values for leisure sector

	Parameter	Normalised value	Reference		Common value range (ref)
TROPOS Leisure					
Visitor centre	Annual visitors	69,277	n.a.	In-project internal data	
	Ticket price	12–25 €	n.a.	In-project internal data	
	CAPEX	3.63 €/visitor	[22–24] Other commercial reports	Article, reports	
	OPEX	8.54 €/visitor year	[22–24] Other commercial reports	Article, reports	
Restaurant	Annual meals	94,133	n.a.	In-project internal data	
	Meal price	15–30 €	n.a.	In-project internal data	
	CAPEX	6.11 €/meal	[22–24] Other commercial reports	Article, reports	
	OPEX	17.11 €/meal year	[22–24] Other commercial reports	Article, reports	
Accommodation	Annual stays	10,220 room nights	n.a.	In-project internal data	
	Stay rate	300 €/night	n.a.	In-project internal data	
	CAPEX	36.68 €/room night	[22–24] Other commercial reports	Article, reports	
	OPEX	43.88 €/room night year	[22–24] Other commercial reports	Article, reports	

3.6 Financial Analysis

In this section, as described in the methodology section, a homogeneous financial analysis of the Ocean of Tomorrow projects is to be developed. This analysis has the objective to test financial performance for all projects under the same assumptions and hypothesis, obtaining indicators that allow comparing the results between projects and comparing them themselves.

Table 3.6 Normalised financial values for container/transport sector

	Parameter	Normalised value	Reference		Common value range (ref)
TROPOS Container terminal					
Container terminal	Throughput	1,000,000 TEU/year	[25]	Article	
	Transhipment	500,000 TEU/year	[25]	Article	
	Price	125 €/TEU	[25]	Article	
	Levelised costs				240–335 €/ TEU [26]
	CAPEX	426.23 €/ TEU	[22–24] Other commercial reports	Article, reports	135–530 €/ TEU [25–27]
	OPEX	26.07 €/ TEU year	[22–24] Other commercial reports	Article, reports	21.8 €/TEU year [27]

The main hypothesis and parameters used in this analysis are:

- CAPEX, OPEX, DECEX and revenues data from previous sections.
- An 8.9% is adopted as uniform discount rate for all projects.
- A 4-year period for construction is considered, plus 20 years for operation and 1 year for decommission.
- Construction costs (CAPEX) are distributed in the 4-year period in 10%, 20%, 40% and 30% percentage.

3.6.1 *TROPOS Commercial Viability*

In Table 3.7 the results of the financial analysis for the four TROPOS designs (Aquaculture, Leisure, Service hub and Container) are presented. In those involving different activities, detailed results are also included. Figure 3.1 provides a graphic summary of the NPV TROPOS design contexts.

Several conclusions arise from the results obtained:

- No design offers financial viability according to their net present values, which result all negative. The most promising design could be "Leisure".
- If financial conditions vary and affect the discount rate, the "Container terminal" design could be profitable if this rate is minimised to 4%. The IRR for the rest of the designs is not reasonable.
- Although no design presents a positive NPV, the production of greater amberjack in the "Aquaculture" design is profitable. While negative, the accommodation module in the "Leisure" design could present easily a positive NPV.

Levelised costs of production for each sector are presented in Table 3.8.

Table 3.7 Results from the financial analysis of TROPOS concepts designs

Design	CAPEX (€ million)	OPEX (€ million/ year)	Revenues (€ million/year)	NPV (€ million	NPV/ CAPEX	Internal rate of return
Aquaculture	−102.74	−43.44	48.35	−69.01	−0.6716	−0.90%
European Seabass	−34.25	−14.48	12.54	−55.84	−1.6303	–
Meagre	−34.25	−14.48	15.87	−25.27	−0.7378	−2.41%
Greater Amberjack	−34.25	−14.48	19.93	12.11	0.3535	12.54%
Leisure	−40.66	−6.26	7.63	−32.57	−0.8010	−4.08%
Visitor centre	−13.61	−2.02	1.73	−17.74	−1.3034	–
Restaurant	−13.66	−2.73	2.82	−14.30	−1.0468	–
Hotel	−13.39	−1.51	3.07	−2.02	−0.1508	8.45%
Service hub	−58.95	−36.27	20.33	−211.92	−3.5949	–
Container terminal	−462.23	−26.07	62.50	−178.28	−0.3856	4.04%

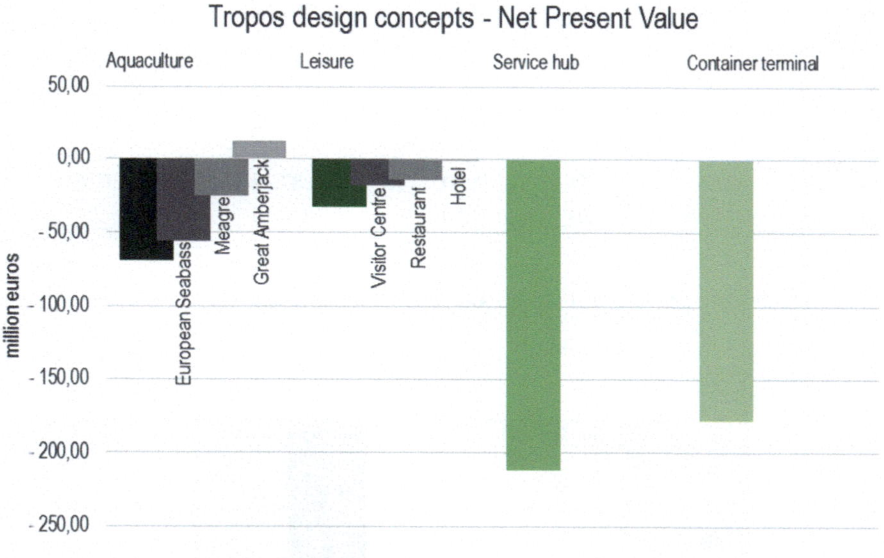

Fig. 3.1 TROPOS design concepts net present value comparison

3.6.2 *MERMAID Commercial Viability*

The results of the financial analysis for the two Mermaid designs analysed (Atlantic and North Sea Sites) are presented in Table 3.9. Detailed results for each sector included in the combinations are also included. Figure 3.2 illustrates the NPV estimations for the Mermaid designs.

Table 3.8 Levelised costs of production for TROPOS design concepts

Design	Unit	Production (units/year)	LCoP (€/unit)
Aquaculture			
European Seabass	Ton of fish	1521	12,240
Meagre	Ton of fish	1521	12,240
Greater Amberjack	Ton of fish	1460	12,751
Leisure			
Visitor centre	Visits	69,277	52.85
Restaurant	Meals	94,133	46.52
Hotel	Night stays	10,220	305.69
Service hub	Services	20,334,129	2.13
Container terminal	TEU	500,000	163.78

Table 3.9 Results from the financial analysis of Mermaid concepts designs

Design	CAPEX (€ million)	OPEX (€ million/year)	Revenues (€ million/year)	NPV (€ million	NPV/ CAPEX	Internal rate of return
Atlantic Site	−2257.45	−53.94	148.14	−1659.19	−0.7349	−2.91%
Wind energy	−1410.90	−33.71	129.75	−695.32	−0.4928	2.24%
Wave energy	−846.54	−20.23	18.39	−963.87	−1.1385	–
North Sea Site	−3037.84	−190.25	437.47	−1129.63	−0.3718	4.15%
Wind energy	−2800.50	−100.00	359.50	−746.90	−0.2667	5.63%
Mussels	−22.58	−32.75	45.00	87.37	3.8693	36.69%
Seaweed	−214.76	−57.50	32.50	−470.10	−2.1889	–

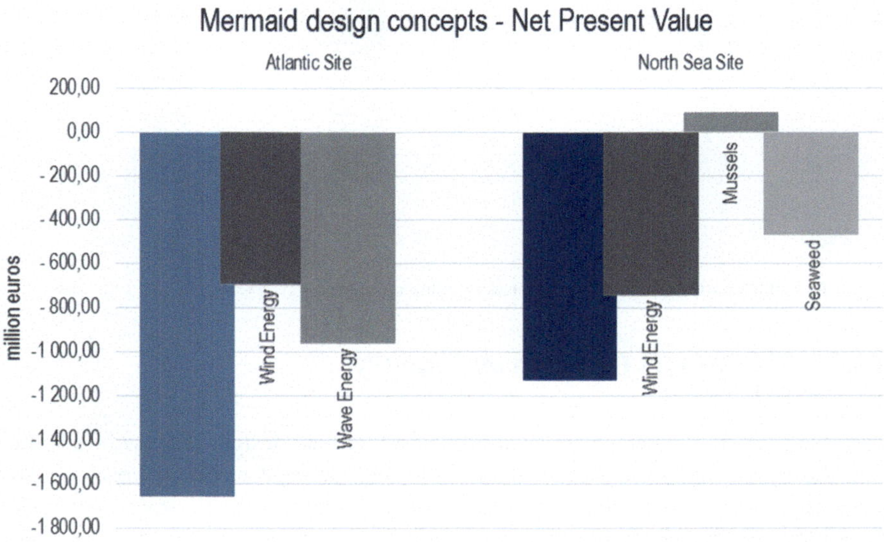

Fig. 3.2 MERMAID design concepts net present value comparison

The results show that:

- None of the design concept shows financial viability according to their NPVs, as all of them result negative.
- Additionally, both designs involve important investments, so their NPVs are negative in a huge dimension, especially when compared with TROPOS designs.
- The results obtained for Mermaid design concepts include in their revenue calculation an important feed-in tariff added to the price of electricity. When or if this subsidy disappears, NPV results will be even more negative. In contrast, any change in policy that rises feed-in tariffs may reinforce the project profitability.

The levelised costs of production for each sector are presented in Table 3.10.

3.6.3 H2Ocean Commercial Viability

For the case of H2Ocean project, a special situation was presented: project deliverables stated the impossible viability of the proposed design. Under their parameter assumptions, the final NPV of the project was 21.6 billion € negative. To allow for a consistent comparison, the results for the joint financial analysis under the proposed values of the parameters are presented here. However, due to the high negative value of the NPV obtained, no further analysis will be performed with this input, nor comparison with the rest of the Oceans of Tomorrow projects (Table 3.11).

Table 3.10 Levelised costs of production for MERMAID design concepts

Design	Unit	Production (units/year)	LCoP (€/unit)
Atlantic Site			
Wind energy	MWh	776,930	264
Wave energy	MWh	110,110	1119
North Sea Site			
Wind energy	MWh	2,600,000	169.51
Mussels	Tons of mussels	48,000	739.51
Seaweed	Tons of seaweed	480,000	174.23

Table 3.11 Results from the financial analysis of H2Ocean concept design

Design	CAPEX (€ million)	OPEX (€ million/year)	Revenues (€ million/year)	NPV (€ million	Internal rate of return
Portugal Site	−11902.13	-455.56	89.70	−16558.42	–
Fish	−3685.38	−166.39	55.85	−5103.76	–
Water	−3328.01	−144.59	0.10	−5022.12	–
Hydrogen/ oxygen	−4888.74	−144.59	33.75	−6432.55	–

3.7 Sensitivity Analysis

The results of the sensitivity analysis performed in the financial model for each of the design concepts are illustrated in the following figures. The percentage showed is the variation of the NPV of the concept when the associated parameter varies by 2.5%. All parameters showing a variation higher than 2.5% should be considered critical for the financial model. The figures also permit identifying the most critical parameters of each concept (Figs. 3.3, 3.4, 3.5, 3.6, 3.7, 3.8 and 3.9).

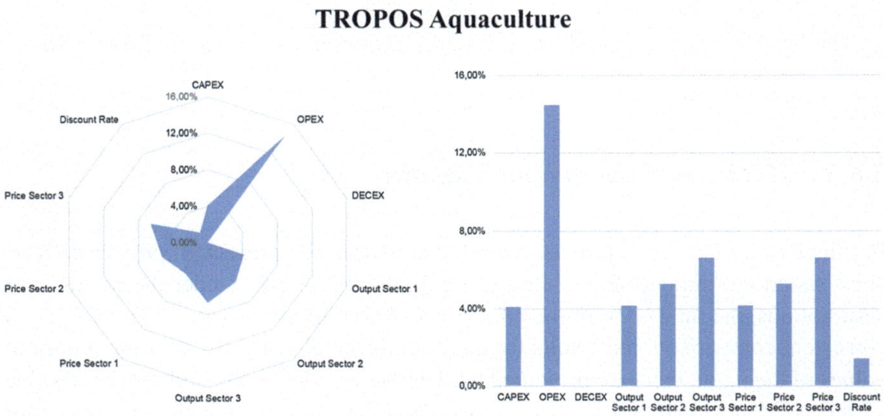

Sector 1 European Seabass
Sector 2 Meagre
Sector 3 Greater Amberjack

Fig. 3.3 TROPOS Aquaculture design concept sensitivity analysis

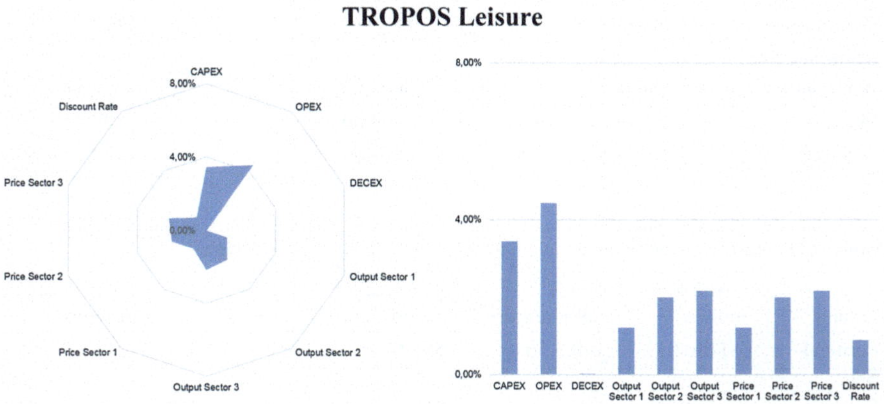

Sector 1 Visitor Centre
Sector 2 Restaurant
Sector 3 Hotel

Fig. 3.4 TROPOS Leisure design concept sensitivity analysis

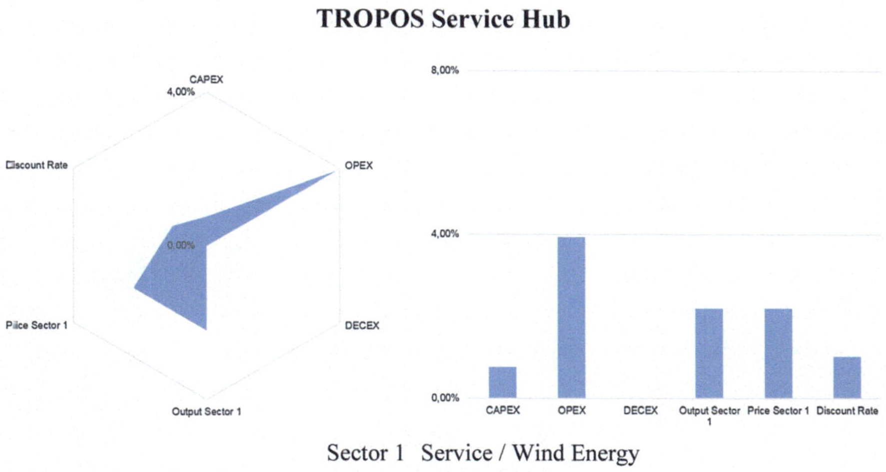

Fig. 3.5 TROPOS Service hub design concept sensitivity analysis

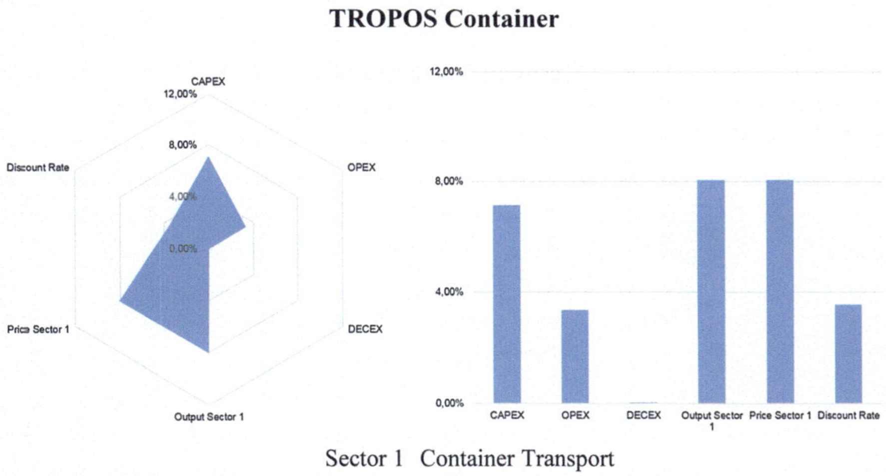

Fig. 3.6 TROPOS Container design concept sensitivity analysis

3.8 Risk Assessment

In this section the results of the probabilistic risk analysis of the Oceans of Tomorrow design concepts studied in the previous sections are presented. The methodology applied follows the recommendations of the European Commission. Table 3.12 summarises the parameters simulated and the probabilistic distributions used. Results are illustrated in Fig. 3.10.

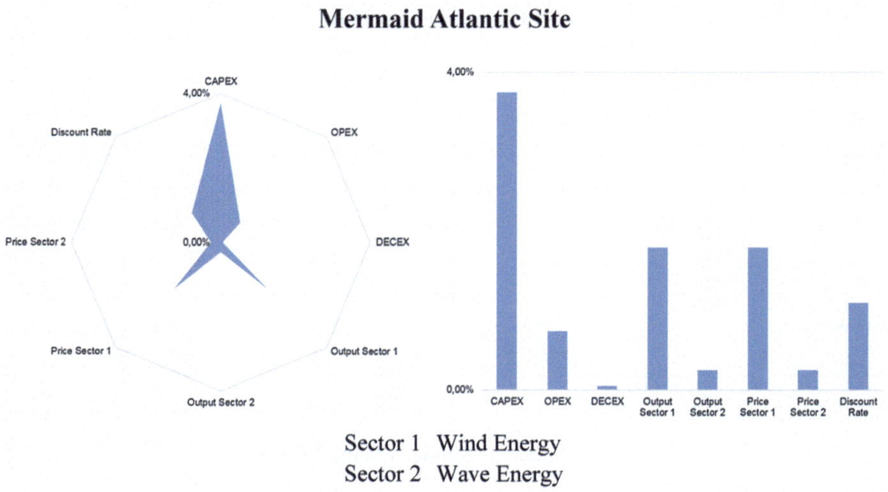

Fig. 3.7 Mermaid Atlantic Site design concept sensitivity analysis

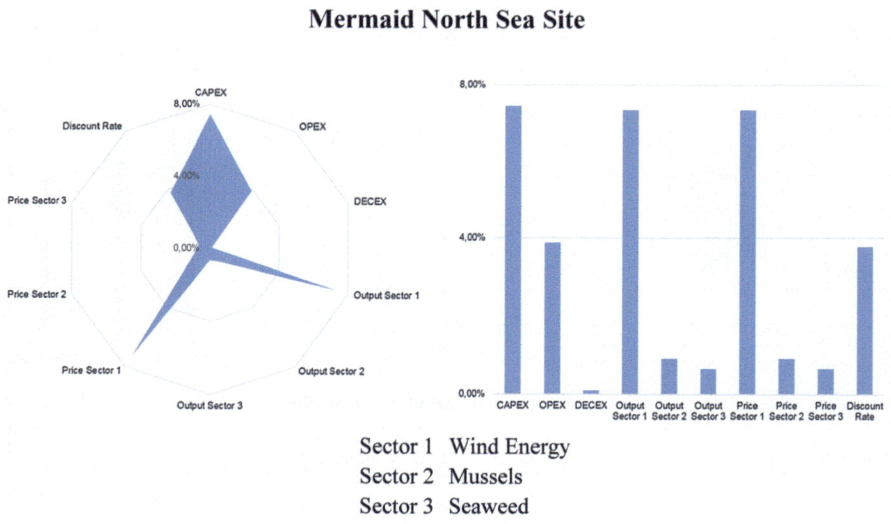

Fig. 3.8 Mermaid North Sea Site design concept sensitivity analysis

Looking at the results, the following points emerge as important to note:

- There is a group of concepts with a similar probabilistic distribution. These are "Aquaculture", "Leisure" and "Container" designs from TROPOS. Although each one varies with respect to NPV values, the maximum-minimum amplitude is similar and so is its maximum probability.

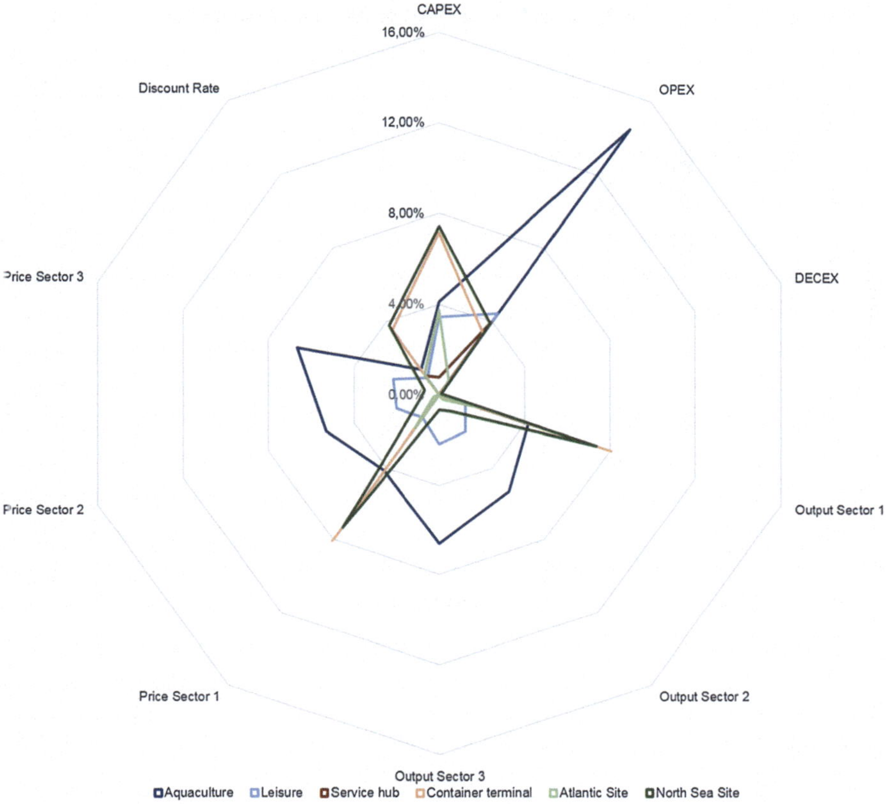

Fig. 3.9 Comparative sensitivity analysis for all design concepts

Table 3.12 Parameters simulated and probabilistic distributions used

Inflow/Outflow	Concept	Distribution
Outflows	Operating costs	Uniform distribution
		Min: 0.85 base OPEX
		Max: 1.15 base OPEX
Inflows	Production	Uniform distribution
		Min: 0.75 production
		Max: 1.00 production
Inflows	Prices	Uniform distribution
		Min: 0.85 base price
		Max: 1.15 base price

- TROPOS “Service hub” presents a very sharp probabilistic distribution. This can be translated into a minimum sensitivity to the parameters analysed and means that the NPV of the design concept is quite defined.
- On the contrary, MERMAID “North Sea Site” and “Atlantic Site” probabilistic distributions are almost flat. This means that the NPVs of both design concepts

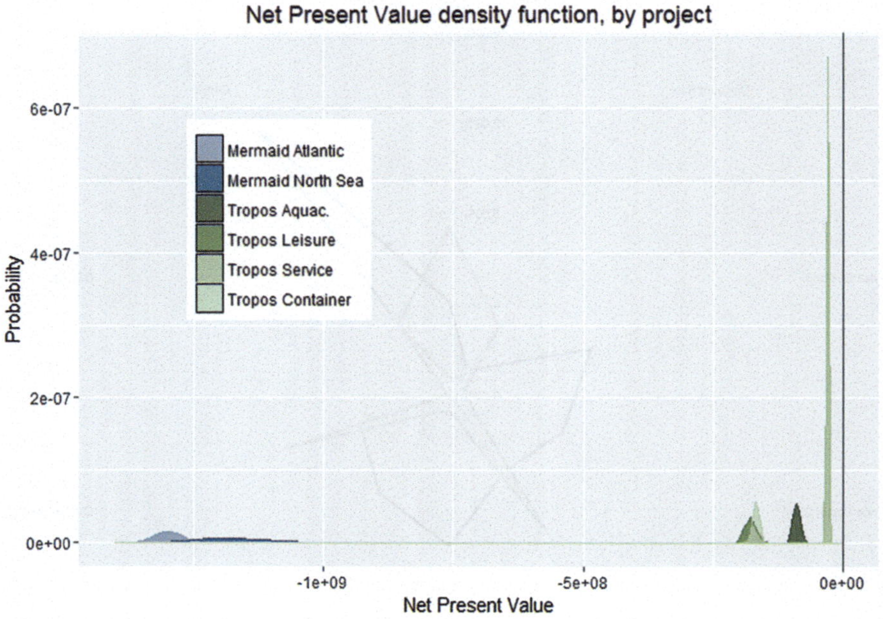

Fig. 3.10 Results from risk analysis of all Ocean of Tomorrow design concepts

accept a wide range of variation, so the viability of the projects is mostly unclear (although always negative so with no financial viability).

3.9 Conclusions

The challenge represented by offshore resources for blue growth economy has been addressed with different strategies, but the results obtained up to now show that the timeline for definitive success is still large. Several reasons have been pointed as explanations such as the lack of adequate technology, required time for technology maturity and adequate knowledge of the effective conditions under which the new business will compete. One of the instruments tested through technology issues in the EU has been the possibility of accelerating expansion through the promotion of sharing resources, experience and space. This possibility has been explored through a set of European-funded projects with different approaches and different maturity levels, focusing on different sectors.

This chapter has presented a transversal analysis of the outputs generated through different projects, trying to clarify the comparison among the existing alternatives, testing them from a standard economic and financial point of view. The results based on the comparison of different projects summarised in this chapter show a homogeneous ranking on the viability of the different alternatives and their business possibilities. The differences in TRL (technology readiness level) explain the different expectations of the projects. What has been proposed as a mere concept idea test cannot be compared with an industrial test under real environmental conditions.

The leadership in offshore activities at present is clearly located in renewable energy and more specifically on offshore wind; hence the most promising combination proposals should be those where this industry appears. However, these proposals show the lowest profitability through our set of projects. This fact calls can be the result of a lower TRL of the combos explored in our sample, to the preeminent role played by huge electrical companies with broad sources of finance, or just to the high level of regulation in this industry, heavily subsidised through feed-in tariffs.

The results presented by aquaculture, seabed, logistics and leisure show more optimistic view in the related projects, but this optimistic analysis does not match with the real investments developed in those sectors in offshore areas; hence the eventual existence of a bias has to be tested in future research. The main sources of uncertainty about the viability of the projects are on one hand the lack of precise knowledge on the operational conditions of the technology under untested conditions, (losses, operational restrictions and external effects between activities) and on the other hand the intrinsic volatility of revenues to be obtained in the future with uncertainty in market conditions, environmental pressures and policy-regulating measures.

Finally, it is important to assume that the initially recognised lack of maturity is still heavily restricting further developments in offshore activities. At the present state of the art, it is unclear that the shared approach creates expectations that it will be a successful strategy.

References

Ali, A. (2005). *Floating transhipment container terminal.* Delft University of Technology.

Bridoux (2008). Algal biomarkers and their metabolites in the lower food web of the Great Lakes (...). Degree of PhD.

Busk, K., & Burton, T. S. (2013). *Financial implications of container terminal automation.* Seaport Group.

Buck, B., Ebeling, M., Michler-Cieluch, T. (2010). Mussel cultivation as a co-use in offshore wind farms. Potential and Economic Feasibility.

Burg, S.V.D., Stuiver, M., Veenstra, F., Bikker, P., Contreras, A. L., Palstra, A., Broeze, J., Jansen, H., Jak, R., Gerritsen, A., Harmsen, P., Kals, J., Blanco, A., Brandenburg, W., Krimpen, M. V., Duijn, A.P. V., Mulder, W., Raamsgink, L.V. (2013) A Triple P review of the feasibility of sustainable offshore seaweed production in the North Sea Burton, T., Sharpe, D., Jenkins, N., & Bossanyi, E. (2001). Wind energy handbook. John Wiley & Sons.

de Alegría, I. M., Martín, J. L., Kortabarria, I., Andreu, J., & Ereño, P. I. (2009). Transmission alternatives for offshore electrical power. *Renewable and Sustainable Energy Reviews, 13* (2009), 1027–1038.

DECC (2013). Electricity Generation Costs.

De Vries, W., Leip, A., Reinds, G. J., Kros, J., Lesschen, J. P., & Bouwman, A. F. (2011). Comparison of land nitrogen budgets for European agriculture by various modeling approaches. *Environmental Pollution, 159*(11), 3254–3268. https://doi.org/10.1016/j.envpol.2011.03.038.

Di Trapani, A. M., Sgroi, F., Testa, R., & Tudisca, S. (2014). Economic comparison between offshore and inshore aquaculture production systems of European sea bass in Italy. *Aquaculture, 434*, 334–339.

Erich, H. (2013). Wind turbines: Fundamentals, technologies, application, economics.

Hopper, T. J. (2016). *The excel user's introduction to R.*

IEA. (2013). Technology roadmap. *Wind Energy.*

Kam, L. E., Leung, P., & Ostrowski, A. C. (2003). Economics of offshore aquaculture of Pacific threadfin (Polydactylus sexfilis) in Hawaii. *Aquaculture, 223*(1), 63–87.

Lagerveld, Sander & Rockmann, C. & Scholl, Michaela & Bartelings, Heleen & Burg, S.W.K. & Jak, Robbert & Jansen, Henrice & Klijnstra, J. & Leopold, Mardik & Poelman, Marnix & Smith, S.R. & Stavenuiter, John & Veenstra, F.A. & Veltman, C. & Westra, C.. (2014). Combining offshore wind energy and large-scale mussel farming: background & technical, ecological and economic considerationsIMARES Wageningen URInstitute for Marine Resources & Ecosystem Studies.

Lipton, D. W., & Kim, D. H. (2007). Assessing the economic viability of offshore aquaculture in Korea: An evaluation based on. Asche, F., & Bjorndal, T. (2011). In *The economics of salmon aquaculture* (Vol. 10). John Wiley & Sons.

Lisac, D., & Muir, J. (2000). Comparative economics of offshore and mariculture facilities. In J. Muir & B. Basurco (Eds.), *Mediterranean offshore mariculture. Zaragoza: CIHEAM* (pp. 203–211) (Options Méditerranéennes: Série B. Etudes et Recherch es; n. 30).

Manwell, J. F., McGowan, J. G., & Rogers, A. L. (2010). *Wind energy explained: Theory, design and application*. Wiley.

Næss-Schmidt, H. S. and Møller, U. (2011). Havvindmøller på vej mod industrialisering: Implikationer for dansk tilgang. Copenhagen Economics, Copenhagen, Denmark.

Ryu, K., Han, H., & Kim, T.-H. (2008). The relationships among overall quick-casual restaurant image, perceived value, customer satisfaction, and behavioral intentions. *International Journal of Hospitality Management, 27*(3), 459–469.

Sartori, D., Catalano, G., Genco, M., Pancotti, C., Sirtori, E., Vignetti, S., & Del Bo, C. (2015). *Guide to cost-benefit analysis of investment projects. Economic appraisal tool for cohesion policy 2014–2020.* Luxembourg: European Commission, Publications Office of the European Union.

Schipper, J., (June 2015). Oral communication by Job Schipper, CEO at Hortimare–Propagating seaweed for a sustainable future.

Virginia Coastal Energy Research Consortium (2010). Virginia Offshore Wind Studies, July 2007 to March 2010, Final Report.

Wiegmans, B. W., Ubbels, B., Rietveld, P., & Nijkamp, P. (2002). Investments in container terminals: Public private partnerships in Europe. *International Journal of Maritime Economics, 4*(1), 1–20.

Chapter 4
Social Acceptance and Socio-economic Effects of Multi-use Offshore Developments: Theory and Applications off the Liuqiu Island

Wenting Chen, Phoebe Koundouri, Osiel González Dávila, Claire Haggett, David Rudolph, Shiau–Yun Lu, Chia–Fa Chi, Jason Yu, Lars Golmen, and Yung–Hsiang (Frank) Ying

Abstract This chapter studies the social acceptance and socio-economic effects associated with the development of multi-use offshore platforms, using a theoretical Green&Blue concept off the Liuqiu Island as a relevant case study. The Green&Blue

---

W. Chen (✉)
Norwegian Institute for Water Research (NIVA), Oslo, Norway
e-mail: wenting.chen@niva.no

P. Koundouri
School of Economics and ReSEES Laboratory, Athens University of Economics and Business, Athens, Greece

United Nations Sustainable Development Solutions Network Europe, Paris, France

EIT Climate–KIC Hub Greece, ATHENA Research and Innovation Center, Athens, Greece

ICRE8: International Centre for Research on the Environment and the Economy, Athens, Greece

O. G. Dávila
Centro de Investigación y Docencia Economícas (CIDE), Mexico City, Mexico

C. Haggett
School of Social and Political Science, University of Edinburgh, Edinburgh, UK

D. Rudolph
Department of Wind Energy, Technical University of Denmark, Kongens Lyngby, Denmark

S.–Y. Lu · C.–F. Chi · J. Yu
National Sun Yat-sen University, Kaohsiung, Taiwan

L. Golmen
Norwegian Institute for Water Research (NIVA), Oslo, Norway

Runde Environmental Centre, Runde, Norway

Tokyo University of Marine Science and Technology, Tokyo, Japan

Y.–H. (F.) Ying
National Taiwan Normal University, Taipei, Taiwan

P. Koundouri (ed.), *The Ocean of Tomorrow*, Environment & Policy 57,
https://doi.org/10.1007/978-3-030-56847-4_4

concept combines the generation of clean energy with the commercial use of biological ocean resources. Social acceptance of such development is evaluated based on a face-to-face survey together with in-depth interviews with local people and tourists who are currently or will be potentially affected by the offshore development. A choice experiment is used to assess the ecosystem services and non-market effects of the offshore development. The social cost-benefit analysis is adopted to synthesise both market and non-market effects. The study finds a generally high support for the offshore development among tourists. The concern mainly focuses on the uncertain environmental impacts and effects on the local fishery industry. Neither locals nor tourists view the energy hub which generates most income and jobs as a very attractive option. The offshore development concept shows a high environmental non-market benefit which amount to 618 million $NT. However, the high investment cost overweighs the positive GDP and environmental gain.

Keywords Offshore development · Multi-use of the sea · Social acceptance · Ecosystem services · Choice experiment · Social costs and benefits

4.1 Introduction[1]

With increasing population growth and intensified competition for space close to the ocean, exploration of the ocean space is attracting new interests, particularly with the development of new technologies (IPCC 2007; Mazor et al. 2014; Wyllie et al. 2017). Sustainable use of ocean is important to reduce the present pressure of human exploitation and obtain enhanced and sustained ecosystem services from ocean (Harris and Tuhumwire 2016). By introducing new technologies and concepts, such as the development of new multi-use offshore platforms, it will be possible to utilise the ocean space in a sustainable manner for future ‘blue growth’. The multiple use of offshore platforms means various functions and productive activities co-existing in the same area, such as shipping, aquaculture, renewable energy and tourism (Chen et al. 2015). The various functions are connected to each other in a sustainable way so as to minimise the impact on each other and to maximise the synergies between them.

Many studies recognised that visual impacts of large offshore constructions affect public acceptance of such constructions (Bishop and Miller 2007; Tsoutsos and Tsouchlaraki 2009; Ladenburg 2008). There are also significant negative welfare effects in terms of environmental degradation from such constructions as highlighted, for example, in Álvarez-Farizo and Hanley (2002) and Busch et al. (2011). The development of multi-use platforms is not only technically challenging and financially costly; it also raises various social and economic issues. In terms of

[1]This work has received funding from the European Union’s 7th Framework Program under grant agreement N° 288710 and N° 288192.

benefits, the platform may bring more income and increase employment for local areas during the construction phase. There are further benefits due to the production of renewable energy and farmed fishes and promotion of tourism that is environment friendly. On the other hand, the construction of such platforms may negatively affect local fauna, disturb existing fishing grounds and affect seascape amenities (Lu et al. 2014). Therefore, it is also important to identify the potential socio-economic effects including both market and non-market effects as well as the individuals whose welfare is likely to be affected before the concept of such platform is put into implementation (Just et al. 2005; Bockstael and Freeman III 2005; Clinch and Murphy 2001). Like other offshore investments, the ultimate success of such concepts will depend on the acceptance and support from local communities and various interest groups (Haggett 2011; Rudolph 2014; Batel et al. 2013; Devine-Wright 2005; Roberts and Boucher 2013).

The paper aims to study the social acceptance and socio-economic effects associated with the development of such offshore platforms off the Liuqiu Island, Taiwan. Social acceptance analysis combines a face-to-face survey together with in-depth interview questions with local people and tourists who are currently or will be potentially affected by the offshore developments. A choice experiment (CE) is deployed to elicit stakeholder preferences in relation to the different platform designs, and to assess the ecosystem services and non-market effects of the platform. To the authors' knowledge, this is the first non-market valuation of multi-use platform and the first one using a CE in this context. The paper then proceeds to use the social cost-benefit analysis to synthesise both market and non-market effects of the platform.

Such offshore investments are likely to become more common because of increased pressures from growing populations and on coastal spaces. This paper provides a guideline and examples for assessing the socio-economic effects of such investments.

The paper is structured as the follows. Section 4.2 details the MUOP concept and its location. Section 4.3 outlines the methodologies used to study the social acceptance and socio-economic effects of the platform. Section 4.4 describes the key findings in terms of the social acceptance of the new offshore development and the socioeconomic effects including the ecosystem service benefits by constructing such a platform.

4.2 New Offshore Multi-use Development: The TROPOS Platform off the Liuqiu Island

This paper reports on some of the results on the TROPOS project (http://www.troposplatform.eu/), which is part of EU FP7 programme 'Ocean of Tomorrow', 'OCEAN Multi-use offshore platforms' (MUOPs), exploring a range of aspects of the development of such platforms. The project focuses on the development of a floating modular multi-use platform system for use in deep waters, with locations in Crete (Greece), Gran Canaria (Spain) and Taiwan (Quevedo et al. 2013). The flexible multi-use platform system will be able to integrate a range of functions

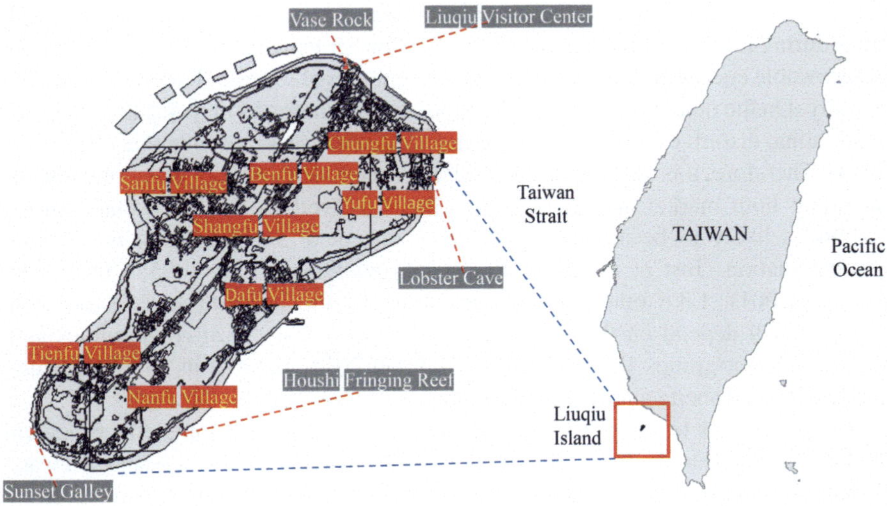

Fig. 4.1 The Liuqiu Island and the villages on the island

from the Transport, Energy, Aquaculture and Leisure sectors (named as TEAL components). The platform aims to reduce land use pressure and make use of the ocean space in a more sustainable manner (Quevedo et al. 2013). The concepts represent a new way to develop the offshore, combining a range of uses, and with the potential to address a range of sustainability related issues. The developed TROPOS platform concepts are all composed of a central unit (CU); different modules, which are integrated into the central unit; and satellite units connected via subsea cables. The platforms are designed around the concept of 'Green & Blue', which means combining the generation of clean energy with the use of biological ocean resources.

One Green & Blue platform scenario is set of off the Liuqiu Island, which is located southwest of Taiwan's main island. The area of the Liuqiu Island is about 6.8 square kilometres, and most of the islanders make their living by fishing. The island has become one of the major tourist destinations and is an important income source for local people (Chen et al. 2015). Figure 4.1 shows the island and villages where surveys were conducted.

The platform concept combines offshore fish and algae aquaculture with Ocean Thermal Energy Conversion (OTEC) for energy supply. All energy needs of the platform will be provided by renewable energy modules. Leisure facilities are also included to accommodate tourists. A total of five modules and a satellite type are designed to fulfil this objective. The components and services included in this platform concept are summarised in Table 4.1.

Aquaculture facilities include fish and algae production as two satellite units. Aquaculture production has the following potential environmental impacts:

(a) Solid and liquid wastes have a major effect on water and sediment quality, benthos, fish and turtles, marine mammals and humans.
(b) Noise and vibrations that affect fish and turtles and marine mammals. The mooring will significantly affect sediment dynamics.

Table 4.1 Components and services of the Green & Blue platform scenario off Liuqiu Island (Chen et al. 2015)

Modules	Components/services	Modules
Fish aquaculture	Fish aquaculture parts of the 30 satellite units	Fish aquaculture module (on CU, operation and control)
Algae aquaculture	Algae aquaculture parts of the 30 satellite units	Algae aquaculture module (on CU, operation and control)
	Bio-refinery (on CU)	Storage (on CU)
Processing plant (CU)	Storage	Processing, packaging, freezing
OTEC plant	Energy (electricity) generation and clean, deepwater supply	
Accommodation	Hotels for tourists and related services	Accommodation for aquaculture staff

(c) Artificial lighting of the fish farm units poses a major impact on marine mammals, birds and bats and fish and turtles.
(d) Escape of fish from the fish cages and the introduction of alien species pose a major impact for plankton, benthos, fish, turtles and potentially the entire ecosystem.

The concept is planned to include a floating closed cycle OTEC plant. Due to heavy structure and autonomy, the OTEC plant is considered as a satellite. OTEC produces constant base load electricity in a turbo generator that is driven by the evaporation/expansion of the working fluid ammonia in a closed circuit. There is significant potential to combine OTEC with aquaculture. The OTEC plant is expected to have moderate effects on the environment, although the stressor heat energy may have an effect on water temperature and the pelagic flora and fauna. Physical stressors owing to potential changes in sea water salinity and water column stratification may also affect the pelagic flora and fauna.

Leisure facilities include accommodation, a restaurant, a sky observation lounge, a garden and a store. Solid and liquid wastes coming from the daily operations of the leisure modules on board of a central unit will most likely have a major effect on marine environment (water, sediment quality, benthos, fish, turtles, marine mammals and humans). To reduce or avoid potential negative impacts of the TROPOS elements on the environment, appropriate mitigation measures are required, in particular for impacts expected to be large or critical for the ecosystem and its receptors.

4.3 Methodology

4.3.1 The Methodology for Assessing Social Acceptance

An integrated multi-use offshore development is a future possibility for the island. However the spatial overlapping with traditional fishing practices and other usages may cause conflict. Evidence shows that local awareness, support and involvement are key elements for developing marine renewable energy (Haggett 2011; Rudolph 2014). Indeed, people's acceptance of a new development is decisive for a project's implementation (Roberts and Boucher 2013). The consideration of social acceptance and local knowledge may thus lead to more competent and well-founded planning decisions.

Social acceptance is a key concept in literature on human geography, sociology and other research on the development of new infrastructure projects (see Batel et al. 2013; Devine-Wright 2005). In their seminal paper on this concept, Wüstenhagen et al. (2007) describe the importance of social acceptance and that it needs to be urgently considered during the implementation of new policies and projects (2007:2683). Understanding social acceptance means exploring the conditions that determine the effective support (and opposition) that any applications receive, and Wüstenhagen et al. (2007) determine that it consists of interrelated 'sociopolitical acceptance', 'community acceptance' and 'market acceptance'.

In this paper, we address in particular the concept of 'community acceptance', which refers to the 'specific acceptance of siting decisions and projects by local stakeholders, particularly residents and local authorities' (Wüstenhagen et al. 2007:2685). Acceptance is determined by a range of interconnected and contextual factors, some of which will relate to the project itself (specifics such as a visual impact; see Haggett 2008); some will relate to the location in which it is planned (such as impact on local wildlife or the local economy; see Rudolph 2014); and some will relate to the process through which the project is being developed (perceived fairness of decision-making processes, the role of public engagement; see Gross 2007; Haggett 2010; Rudolph et al. 2015). All of this matters for a variety of reasons. As Yearley et al. (2003) document, understanding social acceptance may be important pragmatically – a project is more likely to be consented if it has public support (Wolsink 2007). But people also may be viewed as citizens who should be involved and engaged about projects that affect them (Bell et al. 2005); and asking local people about their local area can help to improve a project by accessing detailed and rich local knowledge and understanding (Aitken 2009; Wynne 2006).

Social acceptance is therefore a critical issue with the development of any new project; and we suggest that this is particularly the case with a very new and novel technological innovation such as a new offshore platform. It is important to understand the views of 'the community' – in Wustenhagen et al.'s terminology – this means local people, key stakeholders, to understand the key local issues and how they might affect perceptions of the new project (Aitken et al. 2016; Pieczka and Escobar 2013).

In this research, social acceptance of the multi-use offshore platform was investigated using a multi-method approach (Teddlie and Tashakkori 2011), which was used to capture a range of different responses to this new concept. The intent was to meaningfully capture the key concerns of stakeholders and also to collect a broader sample of information from local people from which generalisable trends could be observed. Understanding the local context is key; and as the platform is being designed in a location dependant on tourism, it was also important to survey tourists to the area. Drawing on best practice from the research methods literature (e.g. Winchester 1999), the approach used here therefore comprised a face-to-face survey with local people and tourists on Liuqiu Island, as well as in-depth interviews with particular stakeholders who are currently or will be potentially affected by offshore developments.

The first part of the multi-method framework applied to reach a wide range of participants was a multiple-choice questionnaire. A large-scale survey was used to give an indicative overview of how the offshore platform is perceived. The central principle of such quantitative studies on social acceptance is to measure and describe relationships and correlations among variables and factors that influence people's acceptance of offshore developments (Roberts and Boucher 2013). In addition, to be able to explore the range of perceptions and potential impacts of the platform in depth, a qualitative interview was used to obtain first-hand information on social realities as they are constructed and presented by various actors, following the approaches of Silverman (2004) and Sin (2003). Rather than solely obtaining structured and quantitative evidence of social acceptance, the qualitative interview can assist in gaining access to reasons behind the different supportive or opposing positions towards the proposed offshore platform (Fielding 2007). Such a research strategy uses people's detailed accounts as a starting point to make sense of the meanings and interpretations, the motives and intentions and arguments that people articulate verbally and that guide and give evidence of their attitudes and behaviour (Blaikie 2010).

The applied interview approach comprised semi-structured interviews, which allowed for some flexibility for the interviewer as well as the interviewee (Fielding 2007). Questions for the interviews were prepared prior to the interview and listed in an interview guide to organise and group themes, issues and questions (May 1997; Fielding 2007). The interview questions made use of insights from previous research on factors likely to determine the acceptance of offshore renewables as discussed above, but also factors relevant to local particularities that may be the object of impacts and concern for people. The sampling of interviewees drew on the suggestions made by previous research on offshore renewables, but also on the findings of the preceding survey (May 1997). Relevant stakeholder groups who may be affected by offshore platforms involve the fishing and shipping community, local leisure industry, coastal communities, local and regional governments, local businesses and tourism as well as other marine users.

4.3.2 *The Methodology for Assessing the Socioeconomic Effects of the Platform*

4.3.2.1 Valuing Ecosystem Services

A key element in the socioeconomic methodology used in the paper is the valuation of ecosystem services to study the impacts of multi-use offshore platforms on the environment and the populations targeted in the TROPOS project. Ecosystem services are widely understood in the literature as the different benefits that humans obtain, directly or indirectly, from natural ecosystems (e.g. Costanza et al. 1997, Daily 1997, De Groot et al. 2002, MEA 2005). The ecosystem services provided by the oceans can be grouped into four main categories: (a) provisioning services such as food and water, (b) regulating services such as climate mitigation, (c) supporting services such as seabed sediment formation and nutrient cycling and (d) cultural services such as recreational, spiritual and other non-material services (MEA 2005). The Marine Strategy Framework Directive (MSFD) (Directive 2008/56/EC) establishes that the economic valuation of offshore projects should follow an ecosystem services approach. It is expected that the multiple tasks to be conducted in the platforms (e.g. energy production, aquaculture and platform-related transport and logistics) will have impacts on marine ecosystem services directly or indirectly. The expected benefits created by platforms include the provisioning services, regulating services and cultural services. They include the production of sustainable food and energy, touristic activities, and several environmental benefits (e.g. improved water quality near coast, climate change mitigation). On the other hand, there are potential negative effects on supporting services. They include the risk of affecting the seabed and the risk to jeopardise populations of fish, mammals and birds in the area. Thus, it is very important to identify and value the different impacts that the proposed structures will have on the ecosystem services. This will help to ensure that all the activities, linked to the design and implementation of the projects, are regulated. Ultimately, the valuation of the ecosystem services will provide useful information to policy makers that can be used to decide whether the project is appropriate for the preservation of a sustainable marine environment and the augmentation of the overall social welfare (Koundouri et al. 2016).

A choice experiment (CE) was conducted in order to identify tourists' and residents' preferences for two different platform designs, design 1 with only aquaculture facilities and design 2 with aquaculture facilities, renewable energy and leisure facilities. The CE method is part of the Total Economic Value framework, which is a standard theoretical approach used for capturing and describing the benefits derived from the different ecosystem services (Defra, 2007). Stated preference methods use structured questionnaires in order to identify the individuals' preferences for a given change in a natural resource or environmental attribute (Champ et al. 2003). Lancaster (1966) explains that any good can be described in terms of its attributes and their levels. Experimental design theory was used to generate different profiles of the platforms in terms of its attributes and their levels. These profiles were then assembled in choice sets and presented to the respondents. Respondents are asked to state their preferences. In this CE, individuals are assumed to choose the design that provides them with the highest utility. The utility function is then used to estimate welfare indicators (willingness to pay (WTP) or willingness to accept (WTA)) based on the levels of attributes (Bennett and Adamowicz 2001; Birol and Koundouri 2008). In this case, the welfare indicators can be understood as the value of changes on the ecosystem services due to the development of the platform.

The random utility theory is the basis for the CE developed in this document, where the utility of a given platform alternative for an individual is a function of the attributes of the platform alternative and of individual socio-economic background features. A second utility relation links the probability of an alternative being chosen to the utility of each alternative. That is, individuals are assumed to choose the alternative yielding the highest utility. An econometric analysis is then conducted using an ordinal logistic model in order to elicit the stakeholder's preferences for different platform designs. The functional forms can be found in Chen et al. (2015). The utility function is then used to estimate welfare indicators (willingness to pay (WTP) and willingness to accept (WTA)) based on the levels of attributes (Bennett and Adamowicz 2001; Birol and Koundouri 2008). In our case the welfare indicators tell the value of ecosystem services change affected by the platform.

4.3.2.2 Social Cost-Benefit Analysis (SCBA)

The social cost-benefit analysis (SCBA) is a technique that assesses the monetary social costs and benefits of an investment project over a time period, in comparison to a well-defined baseline alternative, in our case the status quo situation. In this way, the social costs and benefits of a platform are evaluated and compared, and the long-run economic efficiency of implementing the project of platform is assessed. The methodology pays close attention to both the financial and socio-economic assessments of the project. The financial assessment includes the estimation of financial costs of the investment and the estimation of project development costs, operation and maintenance costs as well as training costs. The socio-economic assessment takes into account all the direct, indirect and induced economic benefits in terms of regional output, income and employment (i.e. market goods), as well as the benefits from positive externalities (i.e. non-market goods) on the environment (through the valuation of the ecosystem services). A project is deemed to be profitable if total social benefits exceed total social costs. Due to the project's expected long-run impacts on the local economy and ecology, its sustainability is to be examined by using the SCBA. The net present value (NPV) or the social net present value (SNPV) of the project is to be estimated using different discount rate schemes (Birol et al. 2010). The NPV/SNPV results reveal whether the net social benefits generated by the investment project of MUOPs are positive and significant well into the future.

4.4 Data Collection and Description

4.4.1 Social Acceptance and Choice Experiment Survey

Survey designs, data collection process and data description are detailed in the TROPOS report (Chen et al. 2015). Here we reiterate the data to brief the readers.

The survey for both social acceptance and choice experiment include a pilot and a final survey. The pilot surveys were used to test the feasibility of the questionnaires before the final full survey. Both the pilot and full surveys for the two were combined together due to limited time and resource of the project. The pilot surveys were conducted on Liuqiu Island between 31 August and 2 September 2014.

4.4.1.1 Survey Design: Social Acceptance

For the social acceptance part, qualitative questions were added after the prestructured questionnaire due to limited time and resources. The questionnaires include questions stakeholders' viewpoints on potential social effects of offshore aquaculture, existing fishing industry, OTEC and tourism. The potential social effects include the effects on health, quality of living environment, economic and material well-being, culture, family and community and institutional legal political equality.

4.4.1.2 Survey Design: Choice Experiment

The choice experiment survey follows the standard five steps, that is, selecting desired attributes, defining levels, choosing the experimental design, constructing choice cards and measuring the preferences. This part of the survey contains questions on respondents' attitudes, 12 choice sets of different levels of attributes and follow-up questions. The attributes describe the potential impacts of the platform on employment and the environment as well as two levels of mitigation and conservation options. The two experimental designs are Design 1 with only aquaculture facilities and Design 2 with aquaculture facilities, renewable energy and leisure facilities (see Annex). There are two environmental impact mitigation levels, acceptable level and optimal level. The acceptable level means the mitigation options will have an acceptable reduction on environmental impacts. The optimal level means the mitigation options will have optimal mitigation options, conservation programs and high visitor satisfaction. For residents, a local tax increase (absolute value per year) is proposed as a payment vehicle. It takes the form of a willingness to pay to avoid environmental damage. This attribute has five levels: (a) 0 euros per year (status quo), (b) 10 euros per year, (c) 20 euros per year, (d) 30 euros per year and (e) 40 euros per year. For tourists, the payment vehicle is a daily tourist tax, that is, an increase of the cost of their holiday in Liuqiu Island per day. The levels for tourist tax were set to (a) 0 euros per day (status quo), (b) 2 euros per day, (c) 4 euros per day, (d) 6 euros per day and (e) 8 euros per day.

4.4.1.3 Data Description

Sample sizes for pilot and full survey are presented in Table 4.2.

All the participants fulfilled the social acceptance part, but not all the participants finished the choice experiment part. Therefore, the sample size of the pilot survey for social acceptance study is different from that of the choice experiment study.

Table 4.2 Sample size for social acceptance in pilot and final full survey (Chen et al. 2015)

	Pilot survey		Final survey
	Social acceptance	Choice experiment	
Residents	22	9	34
Tourists	26	11	118
Total	48	20	152

Table 4.3 Sample distribution among villages (Chen et al. 2015)

			Survey	
Village	Population	Proportion	Optimal number	Actual number
Shangfu	1926	16%	5	5
Dafu	1902	16%	5	5
Chungfu	1364	11%	4	4
Tienfu	1213	10%	4	4
Benfu	2402	20%	7	7
Sanfu	897	7%	3	2
Nanfu	1333	11%	3	3
Yufu	1114	9%	3	4
Total	12,151	100%	34	34

Respondents from the pilot survey complained that the questions on choice modelling were too complicated and too lengthy. The questionnaires used in the final full survey were thus modified for easier completion. Modifications include a question on annual income and adding a map with administrative boundaries of the Liuqiu Island to facilitate respondents to link their marine activities to the platform. Both local residents and tourists were approached randomly in the full survey.

The full survey was carried out in a face-to-face manner between 8 and 16 November on the island. There were 152 participants in total with 118 tourists and 34 residents. Each of the interviewees was shown 12 choice cards and was asked to do the following ranking: 1st = most preferred, 2nd = residual and 3rd = least preferred. The sample size of tourists and residents were arranged in proportion to the size of local population and average monthly tourists in 2013 as required by the choice experiment study. The respondents cover eight villages and five major scenic areas (Table 4.3). The number of respondents from each village was decided according to the population distribution of the Liuqiu Island in 2014.

Table 4.4 GDP impacts and the total multiplier effects of CAPEX and OPEX for the Green & Blue concept in Taiwan

	Cost (million $NT)		GDP impacts (million $NT)	Employment impact (FTE)	Multiplier effect Per annum
Leisure	CAPEX	1767	670	547	1.91
	OPEX (annual)	179	189	104	2.19
Aquaculture	CAPEX	2759,21	955	753	1.83
	OPEX (annual)	1565	672	580	1.85
Energy	CAPEX	3198	1118	878	1.95
	OPEX (annual)	1448	542	391	1.72
Total monetary Value	CAPEX	7724	2743	2178	–
	OPEX (annual)	3192	1403	1075	–

4.4.2 Data on Social Costs and Benefits

Table 4.4 shows the CAPEX, OPEX and the GDP impacts, the approximated employment effects[2] and the total multiplier effects of the investments for the Green & Blue concept of Taiwan. The data are cited from TROPOS (2014) and are calculated by using regional input and output model. The table shows that the GDP impact for the leisure module is the lowest among the three modules. The GDP impact for energy hub is the highest (TROPOS 2014).

The data for calculating social benefits and costs include the market benefits (i.e. effects on GDP[3]) when constructing such a platform and during operational phase, and the non-market benefits (i.e. environmental benefits), investment costs (CAPEX) and operational costs (OPEX). The interest rate used in the baseline study is set at 4%, and life span for the project is 20 years. Table 4.5 provides the variable explanation and values used in the analysis.

[2]The employment effect is estimated according to contribution of GDP to employment in Gran Canaria. The assumption is made due to lack of data in Taiwan.

[3]The GDP effects of CAPEX and OPEX are calculated by using regional input and output model. Details refer to TROPOS (2014).

Table 4.5 Variables explanation and values used in the social cost and benefit analysis

Variable			Unit	Value for each variable	Source
B_t	Market benefit	GDP impact (OPEX aquaculture + leisure + energy)	Million $NT	1403	TROPOS (2014)
	Environmental benefit/non-market benefit	Aquaculture +leisure + energy	Million $NT	618.25	Choice experiment estimates
C_t		OPEX (aquaculture +leisure + energy)	Million $NT	3192	TROPOS (2014)
R	Interest rate			0.04	
T	Project life span		Year	20	
I	CAPEX	Aquaculture + leisure + energy	Million $NT	7724	TROPOS (2014)
GDP_{impact}	GDP impact of CAPEX	Aquaculture + leisure + energy	Million $NT	2743	TROPOS (2014)

4.5 Results for a TROPOS Multi-use Platform in Taiwan

4.5.1 Social Acceptance

A key finding from data collection was a general lack of awareness about the proposed offshore platform project. The high degree of unawareness, particularly among the local population and tourists, may indicate poor project public relations activities and points to the need to improve local stakeholder involvement in the project planning. Awareness raising can help gain support and legitimise the project, to address potential concerns and to embed the project in the local context.

Despite the relative unawareness of the project,[4] the data indicate that the majority of participants initially support such a project, or they stated that they had not formed an opinion yet. 40% of residents and 73% of tourists choose more likely to support the project when being confronted with the proposal of the Green & Blue concept.

Despite the general acceptance of such a project, a number of concerns were raised. These concerns were predominantly related to environmental impacts and unclear effects on local fishing and fish processing industries. Key concerns seem to be based on the concept and use of the platform, which overlaps with and may thus destabilise existing industries. Other issues that challenge the acceptance of the

[4]The general awareness of the project among stakeholders may be explained by the fact that the project is still at a hypothetical research phase, not at the planning phase.

Table 4.6 Estimated willingness to pay per tourist per day for Design 1 and Design 2 with acceptable environmental mitigation level

	Coef.	Std. err.	z	P > |z|	[95% conf. interval]	
WTP for Design 1	−86.52	33.93	−2.55	0.01	−153.03	−20.00
WTP for Design 2	53.66	23.56	2.28	0.02	7.48	99.85

project include uncertain environmental impacts and adverse effects caused by the construction of the platform.

Negative impacts are contrasted with likely benefits for the tourist sector, which is another crucial economic driver for the Liuqiu Island. Benefits for tourism businesses are predicted to result in an increase of income and generation of jobs, due to increasing tourist numbers and a boost to the public image of the local area. The generally critical views on potential impacts on the environment and fishing industries seem to have been strengthened by the economic foundation of the area.

Our research has shown that people are mainly concerned with environmental impacts and potential disruptions to the existing fishing sector. This points to the need for thorough consideration of these concerns in the planning and development process in order to get local citizens on board with the project, to legitimise it and to integrate the platform more effectively in the local context.

4.5.2 Ecosystem Services Value

The main objective was to identify stakeholder preferences, the willingness to pay, for two different platform designs for the Liuqiu Island. In the study we only focus on the use value, which includes the direct use value (provision services and cultural services) and the indirect use value (regulating services). We distinguish the local residents and tourists in the choice experiment survey as they have very different preferences. In the paper we only include the results from the tourists as the sample size for local residents is too small to provide valid and robust estimates. Among tourists there are about 41.18% respondents who preferred the status quo option and 40.96% who preferred Design 2 which combines aquaculture, OTEC and leisure facilities. Design 2 is preferred over Design 1 which includes only aquaculture facilities. This may be due to the fact that Design 2 offers leisure facilities that could be used by the tourists.

The results also indicate that the higher the tax, the less likely that the option will be selected as the most preferred by tourists. Table 4.6 shows the estimated daily willingness to pay per tourist for Design 1 and Design 2 with acceptable environmental mitigation level. If renewable energy and leisure facilities are developed (Design 2), a willingness to pay for a tax per day is estimated to be NT$ 53.66. The total willingness to pay for the whole of Taiwan would amount to NT$ 618.25 million. It should be noted that with only aquaculture facilities presented (Design 1), tourists require a compensation instead. This compensation was estimated at NT \$86.52.

4.5.3 Social Costs and Benefits

In this section we compare the GDP benefits from the investments and the environmental benefits with the capital and investment costs. By including the non-market value of environmental effects, NPV could be regarded as social net present value, i.e. SNPV. SNPV follows Eq. (4.1):

$$SNPV = -I + GDP_{\mathrm{impact}} + \sum_{t=1}^{20} (B_t - C_t)/(1+r)^t \tag{4.1}$$

where I is the total monetary capital cost. GDP_{impact} stands for the GDP impact of CAPEX, B_t is the annual social benefit of the platform, C_t is annual social cost of the platform which is equivalent to the financial cost here, t is the life span of the platform, and r is the annual interest rate. The project is accepted if SNPV >0; otherwise it can be rejected. When the SNPV is used to decide which project alternative should be chosen, the project with the highest positive SNPV should be preferred.

Table 4.7 shows the estimation of SNPV at an annual base for the multifunctional platform with and without considering the non-market value of environmental effects. As presented in the table, the SNPVs are negative no matter whether the non-market value of environmental effect is included or not, mainly due to the huge investment costs (Chen et al. 2015). Sensitivity analysis is carried out with different discount rate and time span for the project. Table 4.8 and Table 4.9 show the results from sensitivity analysis with 5% interest rate, and with 70-year time span for the platform respectively. The SNPV are negative in both sensitivity analyses.

Table 4.7 SNPV for the platform with aquaculture, OTEC and leisure: with and without non-market value of environmental effects, r = 4% and t = 20 (unit: million \$NT)

	Green & Blue concept: Aquaculture + OTEC + leisure	
	Without non-market value of environmental effects	With non-market value of environmental effects
SNPV	−29294.41	−20892.21

Table 4.8 Sensitivity analysis SNPV with WTP for sustainable development r = 5% and t = 20 (unit: million \$NT)

	Aquaculture + OTEC + leisure	
	Without WTP for sustainable development	With WTP for sustainable development
SNPV	−27276.21	−19571.46

Table 4.9 Sensitivity analysis SNPV with WTP for sustainable development r = 4% and t = 70 (unit: million $NT)

	Aquaculture + OTEC + leisure	
	Without WTP for sustainable development	With WTP for sustainable development
SNPV	−46834.10	−32370.48

4.6 Conclusion and Discussion

The study found a general lack of awareness about the suggested offshore platform project off the Liuqiu Island. Our results highlight the concerns about environmental effects of the platform and the unknown effects on existing industries, such as fishing and fish processing. Other concerns from respondents include questions about who will be responsible for the platform's construction and operation. Most residents were worried the platform could be destroyed by storm waves. Sustainable tourism is regarded as positive by local stakeholders. The findings suggest that current project public relations activities were not proactive and indicate the need to improve involvement of local people and existing industries before the project could potentially be carried out. Sie et al. (2018) use a group model building approach to improve the stakeholders' understanding of the complex systems and multi-functions of the TROPOS platform. Under the approach the stakeholders are able to discuss and present abstract opinions, expected system behaviours and important factors in a macro-system structure (Sie et al. 2018). Their general findings are in lieu with our results from social acceptance study. They also find, for example, that stakeholders concern about the negative side of tourism including potential decline of quality of service, traffic jams and negative effects on marine ecological system. The stakeholders also agreed that the platform will enhance fishery ecology nearby the island when the platform plansan integrated multi-trophic aquaculture.

Our findings further show that costs and environment effects are factors that influence stated preferences to a high degree. The presence of energy and leisure facilities moderately affected preferences, and GDP effects and job creation were not deemed very important factors when preferences were stated. Tourists would support the installation of the platform only if renewable energy and leisure facilities were provided. The total willingness to pay for the whole of Taiwan would amount to NT$ 618.25 million. Due to the large investment costs of the platform, the social SNPV is negative, even if the environmental benefits are considered.

In the end a couple of issues need to pay attention to but have not mentioned explicitly in our study. The SNPV analysis implies that the GDP impact, capital investment and operational costs as well as the ecosystem service values are additive and mutually exclusive (Munda 1996). GDP impact is calculated by input-output model and considered only the economic sectors related to platform investments during construction and operation.Therefore, it does not overlap with the production service and cultural service values obtained from choice experiment. The sectors covered by input-output analysis can be found in TROPOS (2014). Our social cost and benefit study assume that the environmental effects have the same time span as the project. This is mainly due to the fact that we cannot distinguish the regulating service value from value of provisioning and cultural service in choice experiment. In cases when environmental effects last much longer than the project life span, different time span should be used in the social cost and benefit analysis. Positive present value of environmental benefits could outweigh the other costs. This is important for environmental goods such as good sea water quality and biodiversity. Finally, our study only focuses on the efficiency. Intra- and intergeneration equities are also important when applying cost-benefit analysis to the environmental issues (Munda 1996). These are issues we will improve in future research.

Acknowledgements The research is a collaboration between the TROPOS project (Grant Agreement No. 288192) (www.tropospatform.eu) and the MERMAID project (Grant Agreement No. 288710) (http://www.mermaidproject.eu/). The TROPOS project and the MERMAIN project were funded by the EU Commission as part of the Seventh Framework programme 'Oceans of Tomorrow' theme, OCEAN.2011-1: Multi-use Offshore Platforms. Funding for Taiwan partners comes from the Ministry of Science and Technology, Taiwan, under the EU FP7 Cooperation Project – Modular Multi-Use Deep Water Offshore Platform Harnessing and Servicing Mediterranean, Subtropical and Tropical Marine and Maritime Resources (TROPOS), Project Number: 101-2923-I-110-001-MY3. Dr. Osiel González Dávila wishes to acknowledge the support granted by Catedras Conacyt, Project 874 'Programa de Estudios Longitudinales, Experimentos y Encuestas para el Análisis de la Pobreza' during the writing up of the final draft. We are grateful to all the participants who are involved in the collaboration.

Annex

Design 1: Aquaculture Facilities

Attributes	Description and economic impacts	Environmental Impacts	Levels	
Design 1: Aquaculture facilities (fish +algae): Satellite unit (not inside the platform)	Fish and algae aquaculture: 1333 FTE positions and GDP impact of NT$ 1660 million (€43.35 million)	**Solid and liquid wastes:** Major effect on water and sediment quality, benthos, fish and turtles, marine mammals and humans	1	Acceptable reduction on environmental impacts
		Noise and vibrations: Fish and turtles and marine mammals, the mooring will significantly affect sediment dynamics		
		Artificial lighting of the fish farm units: Pose a major impact on marine mammals, birds and bats and fish and turtles	2	Optimal levels of conservation andhigh visitor satisfaction
		Escape of fish from the fish cages and the introduction of alien species: Major impact for plankton, benthos and fish and turtles		

Design 2: Aquaculture Facilities + Renewable Energy: OTEC plant + Leisure Facilities

Attributes	Description and economic impacts	Environmental impacts	Mitigation levels	
Design 2: Aquaculture facilities (fish+algae): Satellite unit (not inside the platform)+ +renewable energy: OTEC plant not inside the platform +leisure facilities (accommodation +food and beverage)	Fish and algae aquaculture: 1333 FTE positions and GDP impact of NT$ 1660 million (€43.35 million)	**Solid and liquid wastes:** Major effect on water and sediment quality, benthos, fish and turtles, marine mammals and humans	1	Acceptable reduction on environmental impacts
		Noise and vibrations: Fish and turtles and marine mammals, the mooring will significantly affect sediment dynamics		
		Artificial lighting of the fish farm units: Pose a major impact on marine mammals, birds and bats and fish and turtles		
		Escape of fish from the fish cages and the introduction of alien species: Major impact for plankton, benthos and fish and turtles	2	Optimal levels of conservation and high visitor satisfaction
	Renewable energy + accommodation, restaurant, sky lounge, garden and store	**Heat energy:** Major effect on water temperature and the pelagic flora and fauna		
		Solid and liquid wastes: Major effect on water and sediment quality, benthos, fish and turtles, marine mammals and humans		

References

Aitken, M. (2009). Wind power planning controversies and the construction of 'expert' and 'lay' knowledges. *Science as Culture, 18*(1), 47–64.

Aitken, M., Haggett, C., & Rudolph, D. (2016). Practices and rationales of community engagement with wind farms: Awareness raising, consultation, empowerment. *Planning Theory and Practice, 17*(4), 557–576.

Álvarez-Farizo, B, & Hanley, N. 2002. Using conjoint analysis to quantify public preferences over the environmental impacts of wind farms. An Example from Spain. *Energy Policy*. http://www.sciencedirect.com/science/article/pii/S0301421501000635

Batel, S., Devine-Wright, P., & Tangeland, T. (2013). Social acceptance of low carbon energy and associated infrastructures: A critical discussion. *Energy Policy, 58*, 1–5.

Bell, D., Gray, T., & Haggett, C. (2005). Policy, participation and the social gap in wind farm siting decisions. *Environmental Politics, 14*(4), 460–477.

Bennett, J., & Adamowicz, V. (2001). Some fundamentals of environmental choice modelling. In J. Bennett & R. Blamey (Eds.), *The choice modelling approach to environmental valuation* (pp. 37–69). Cheltenham: Edward Elgar Publishing Limited.

Birol, E., & Koundouri, P. (2008). *Choice experiments in Europe: Economic theory and applications*. Northampton: Edward-Elgar Publishing, Inc. Wally Oates and Henk Folmer's 'New Horizons in Environmental Economics' series, ISBN: 9781845427252 (337pages).

Birol, E., Koundouri, P., & Kountouris, Y. (2010). Assessing the economic viability of alternative water resources in water-scarce regions: Combining economic valuation, cost-benefit analysis and discounting. *Ecological Economics, Elsevier, 69*(4), 839–847.

Bishop, I. D., & Miller, D. R. (2007). Visual assessment of off-shore wind turbines: The influence of distance, contrast, movement and social variables. *Renewable Energy*. http://www.sciencedirect.com/science/article/pii/S0960148106000838

Blaikie, N. (2010). *Designing social research* (2nd ed.). Cambridge: Polity Press.

Bockstael, N. E., & Freeman, A. M., III. (2005). Welfare theory and valuation. *Handbook of Environmental Economics, 2*, 517–570.

Busch, M., Gee, K., Burkhard, B., Lange, M., & Stelljes, N. (2011). Conceptualizing the link between marine ecosystem services and human Well-being: The case of offshore wind farming. *International Journal of Biodiversity Science, Ecosystem Services & Management, 7*(3), 190–203. https://doi.org/10.1080/21513732.2011.618465.

Champ, P. A., Boyle, K. J., & Brown, T. C. (Eds.). (2003). *A primer on nonmarket valuation*. Dordrecht: Kluwer Academic Publishers.

Chen, W; Koundouri, P., Dávila, O.G. Souliotis, Y, Haggett, C; Ruldoph, D; Ying, F. Lu, S. Y, Chi, C. Lin, J. Li, S. Mintenbeck, K., Golmen, L., Quevedo, E., & Grito, J. H. (2015). *D6.6 A framework for describing the social impact with concrete examples that apply for Green and Blue Concept in Taiwan- a joint report between EU FP7 TROPOS and EU FP7 MERMAID, EU FP7 TROPOS project*. Publication Office of the European Union: Luxembourg. http://www.troposplatform.eu/Deliverables-Media

Clinch, J., & Murphy, A. (2001). Modelling winners and losers in contingent valuation of public goods: Appropriate welfare measures and econometric analysis. *The Economic Journal, 111*(470), 420–443.

Costanza, R., d'Arge, R., De Groot, R., Faber, S., Grasso, M., Hannon, B., et al. (1997). The value of the world's ecosystem services and natural capital. *Nature, 387*, 253–260.

Daily, G. (1997). *Nature's services: Societal dependence on natural ecosystems*. Washington, DC: Island Press.

Defra (2007).An introductory guide to valuing ecosystems. PB12852, Department of Environment, Food and Rural Affairs, London. http://www.defra.gov.uk/environment/policy/naturalenviron/documents/eco-valuing.pdf.

De Groot, R. S., Wilson, M. A., & Boumans, R. M. (2002). A typology for the classification, description and valuation of ecosystem functions, goods and services. *Ecological Economics, 41*(3), 393–408.

Devine-Wright, P. (2005). Beyond NIMBYism: Towards an integrated framework for understanding public perceptions of wind energy. *Wind Energy, 8*, 125–139.

Fielding, N. (2007). Qualitative interviewing. In N. Gilbert (Ed.), *Researching social life*. London: Sage.

Gross, C. (2007). Community perspectives of wind energy in Australia: The application of a justice and community fairness framework to increase social acceptance. *Energy Policy, 35*(5), 2727–2736.

Haggett, C. (2008). Over the sea and far away? A consideration of the planning, politics, and public perceptions of offshore wind farms. *Journal of Environmental Policy and Planning, 10*(3), 289–306.

Haggett, C. (2010). The principles, procedures, and pitfalls of public engagement in decision-making about renewable energy. In P. Devine-Wright (Ed.), *Renewable energy and the public*. London: Earthscan.

Haggett, C. (2011). Understanding public responses to offshore wind power. *Energy Policy, 39*(2), 503–510.

Harris, P., & Tuhumwire, J. (2016). Chapter 1: UN World Ocean assessment. In: Introduction – Planet, oceans and life. http://www.worldoceanassessment.org/

IPCC. (2007). *Intergovernmental Panel on Climate Change, Climate Change 2007: Working group II: Impacts, adaptation and vulnerability*. IPCC Fourth Assessment Report: Climate Change 2007 [online]. [Cited November 28, 2017]. https://www.ipcc.ch/publications_and_data/ar4/wg2/en/ch6s6-2-2.html

Just, R. E., Hueth, D. L., & Schmitz, A. (2005). *The welfare economics of public policy: A practical approach to project and policy evaluation*. Edward Elgar Publishing.

Koundouri, P., Souliotis, I., & Giannouli, A. (2016). An integrated approach for sustainable environmental and socio-economic development using offshore infrastructure. In M. Erdoğdu, T. Arun, & I. Ahmad (Eds.), *Handbook of research on green economic development initiatives and strategies* (pp. 44–64). Hershey: HabibIGI Global.

Ladenburg, J. (2008). Attitudes towards on-land and offshore wind power development in Denmark; choice of development strategy. *Renewable Energy* http://www.sciencedirect.com/science/article/pii/S0960148107000201

Lancaster, K. J. (1966). A new approach to consumer theory. *The Journal of Political Economy*, 132–157.

Lu, S.-Y., Yu, J., Wesnigk, E., Delory, E., Quevedo, J., Hernández, O., Llinás, L., Golmen, N., Papandroulakis, & Anastasiadis, P. (2014). Environmental aspects of designing multi-purpose offshore platforms in the scope of the FP7 TROPOS Project. *Proceedings OCEANS 2014 Taipei, IEEE Explore*. https://doi.org/10.1109/OCEANS-TAIPEI.2014.6964306, 8 p.

Marine Strategy Framework Directive (MSFD). (2008). Directive 2008/56/EC of the European Parliament and of the Council of 17 June 2008 establishing a framework for community action in the field of marine environmental policy. *Official Journal of the European Union, L 164* (25.6.2008), 19–40.

May, T. (1997). *Social research: Issues, methods and process* (2nd ed.). Buckingham: Open University Press.

Mazor, T., Possingham, H. P., Edelist, D., Brokovich, E., & Kark, S. (2014). The Crowded Sea: Incorporating multiple marine activities in conservation plans can significantly alter spatial priorities. *PLoS One, 9*(8), e104489. https://doi.org/10.1371/journal.pone.0104489.

Millennium Ecosystem Assessment (MEA). (2005). *Ecosystems and human Well-being: Synthesis*. Washington, DC: Island Press.

Munda, G. (1996). Cost-benefit analysis in integrated environmental assessment: Some methodological issues. *Ecological Economics, 19*, 157–168.

Pieczka, M., & Escobar, O. (2013). Dialogue and science: Innovation in policy-making and the discourse of public engagement in the UK. *Science and Public Policy, 40*(1), 113–126.

Quevedo, E., Cartón, M., Delory, E., Castro, A., Hernández, J., Llinás, O., Bard, J., de Lara, J., Jeffrey, H., Ingram, D., Papandroulakis, N., Anastasiadis, P., & Wesnigk, J. (2013). *Multi-use offshore platform configurations in the scope of the FP7 TROPOS project*. Bergen: MTS IEEE OCEANS.

Roberts, T., & Boucher, P. (2013). Methodologies for understanding low-carbon controversies. In T. Roberts, P. Upham, S. Mander, C. McLachlan, P. Boucher, C. Gough, & D. Abi Ghanem (Eds.), *Low-carbon energy controversies* (pp. 44–60). London: Routledge.

Rudolph, D. (2014). The resurgent conflict between offshore wind farms and tourism: Underlying storylines. *Scottish Geographical Journal, 130*(3), 168–187.

Rudolph, D., Haggett, C. & Aitken, M. (2015). *Community benefits from offshore renewables: good practice review*. www.climatexchange.org.uk

Sie, Y. T., Château, P. A., Chang, Y. C., & Lu, S. Y. (2018). Stakeholders opinions on multi-use deep water offshore platform in Hsiao-Liu-Chiu, Taiwan. *International Journal of Environmental Research and Public Health, 15*(2), 281. https://doi.org/10.3390/ijerph15020281.

Silverman, D. (Ed.). (2004). *Qualitative research: Theory, method and practice* (2nd ed.). London: Sage.

Sin, C. H. (2003). Interviewing in 'place': The socio-spatial construction of interview data. *Area, 35*(3), 305–312.

Teddlie, C., & Tashakkori, A. (2011). Mixed methods research. Contemporary issues in an emerging field. In N. K. Denzin & Y. S. Lincoln (Eds.), *The SAGE handbook of qualitative research* (4th ed., pp. 285–299). Thousand Oaks: Sage.

TROPOS. (2014). An assessment of the economic impact, on local and regional economies, of the large scale deployment. EU FP7 TROPOS project Deliverable 5.2. http://www.tropositplatform.eu/Deliverables-Media/Project-Deliverables/(offset)/15

Tsoutsos, T, & Tsouchlaraki, A. (2009). Visual impact evaluation of a Wind Park in a Greek Island. *Applied Energy*. http://www.sciencedirect.com/science/article/pii/S0306261908002079.

Winchester, H. P. M. (1999). Interviews and questionnaires as mixed methods in population geography. *The Professional Geographer, 51*, 60–67.

Wolsink, M. (2007). Planning of renewables schemes: Deliberative and fair decision-making on landscape issues instead of reproachful accusations of non-cooperation. *Energy Policy, 35*(5), 2692–2704.

Wüstenhagen, R., Wolsink, M., & Bürer, M. J. (2007). Social acceptance of renewable energy innovation: An introduction to the concept. *Energy Policy, 35*(5), 2683–2691.

Wyllie, M., Newport, A., & Mastrangelo, C. (2017). *The benefits and limits of FPSO standardisation*. Offshore technology conference, 1–4 May, Houston, Texas, USA. https://doi.org/10.4043/27545-MS

Wynne, B. (2006). Public engagement as a means of restoring public trust in science–hitting the notes, but missing the music? *Public Health Genomics, 9*(3), 211–220.

Yearley, S., Cinderby, S., Forrester, J., Bailey, P., & Rosen, P. (2003). Participatory modelling and the local governance of the politics of UK air pollution: A three city case study. *Environmental Values, 12*(2), 247–262.

Chapter 5
An Interdisciplinary Web-Based Decision Support System for Socio-economic Assessment of Marine Investments: The MERMAID Project

Evita Mailli, Petros Xepapadeas, and Phoebe Koundouri

Abstract This chapter (this work has received funding from the European Union's seventh Framework Program under grant agreement N° 288,710) describes a decision support tool that was designed and developed for the MERMAID project (EU-FP7), which has already indicated in this book developed concepts for next-generation offshore platforms for multi-use of ocean space for energy extraction, aquaculture, and platform-related transport. Specifically, it evaluated the potential and challenges of building multi-use offshore platforms (MUOPs). The MERMAID project considers four offshore study sites for multi-use offshore platforms, Atlantic Ocean site, Wadden-North Sea site, Baltic Sea site, and Mediterranean Sea site. Each site is considered in terms of its available resources and unique features. This tool was part of the framework for assessing the socio-economic impact of MUOPs and, as such, utilized web and data analytics state-of-the-art technologies in order to provide researchers with a framework for evaluating feasibility and potential of each MUOP's proposed design and location.

E. Mailli
Department of Informatics at the National and Kapodistrian University of Athens, Athens, Greece

ICRE8: International Centre for Research on the Environment and the Economy, Athens, Greece

P. Xepapadeas
ICRE8: International Centre for Research on the Environment and the Economy, Athens, Greece

Department of Management Science and Technology at the Athens University of Economics and Business, Athens, Greece

P. Koundouri (✉)
School of Economics and ReSEES Laboratory, Athens University of Economics and Business, Athens, Greece

United Nations Sustainable Development Solutions Network Europe, Paris, France

EIT Climate-KIC Hub Greece, ATHENA Research and Innovation Center, Athens, Greece

ICRE8: International Centre for Research on the Environment and the Economy, Athens, Greece
e-mail: pkoundouri@aueb.gr

P. Koundouri (ed.), *The Ocean of Tomorrow*, Environment & Policy 57,
https://doi.org/10.1007/978-3-030-56847-4_5

Keywords MERMAID · Multi-use offshore platforms · Socio-economic assessment · Energy extraction · Aquaculture · Transport · Web-based tool

5.1 Introduction

This chapter describes a decision support tool that was designed and developed for the MERMAID project (EU-FP7), which has already indicated in this book developed concepts for next-generation offshore platforms for multi-use of ocean space for energy extraction, aquaculture, and platform-related transport. Specifically, it evaluated the potential and challenges of building multi-use offshore platforms (MUOPs) (MERMAID 2014). The MERMAID project considers four offshore study sites for multi-use offshore platforms, Atlantic Ocean site, Wadden-North Sea site, Baltic Sea site, and Mediterranean Sea site. Each site is considered in terms of its available resources and unique features. This tool was part of the framework for assessing the socio-economic impact of MUOPs and, as such, utilized web and data analytics state of the art technologies in order to provide researchers with a framework for evaluating feasibility and potential of each MUOP's proposed design and location (Koundouri 2017).

Firstly, we will present the purpose, as well as an overview of the design and information flow in the tool. We will then describe our methodology and implementation of this tool, emphasizing the structure and underlying assumptions based on project MERMAID's particular characteristics. We will then briefly describe the tool's technology and architecture. Finally, we will outline our conclusions and highlight additional features that we think can further enhance this tool, as well as similar tools.

5.2 Overview of the Tool

5.2.1 *Purpose and Underlying Structure*

The assessment tool developed in MERMAID project is an electronic decision support system that applies an Integrated Socio-Economic Assessment methodology for identifying the optimal design and combination of provision of services, of multi-use offshore platforms. The e-tool integrates legal, institutional, technological, financial, socio-economic, and environmental parameters. It provides the user with questionnaires for technical and legal feasibility assessment, as well as financial, economic, and environmental impact assessment. Based on the user's answers, it advices on the feasibility and sustainability of a potential offshore platform investment. The MERMAID decision support system (DSS) uses data provided by the users in a predetermined format, while it enables the user to evaluate the risk of the investment for a particular location site by comparing different net present value rates.

5.2.2 *Workflow*

The MERMAID project defines four sites, Mediterranean, North, Baltic, and Atlantic, which comprise the four case studies of the assessment tool. Each site is modeled differently in the tool due to variability in area characteristics that lead to different ecological, engineering, socio-economic, and financial concerns. The MERMAID assessment tool consists of the following five parts:

Part 1: Technical and Legal Feasibility Assessment In this part, the user will have to provide answers to a list of predefined questions, such as the ones shown in Fig. 5.1. The answers are quantified and inserted into the tool, which outputs a message on the environmental possibility of the platform. The first questions represent the main aspects that need to be taken into account for the legal and technical feasibility. The tool quantifies the answers and feeds them into an algorithm that displays a message of whether the user may continue with the rest of the process, or a message could be shown based on the unmet technical or legal constraints, i.e., if the answers to the last questions are negative (Fig. 5.1).

Part 2: Environmental Impact Assessment Regarding the Environmental Impact Assessment (EIA), the users are asked to identify all significantly positive and/or negative environmental impacts (at local, regional, and global levels). Also, they are asked if there is an EIA available for similar project(s) in the region. The set of risks identified for this section refer to the uncertainty about climate change and other environmental parameters, the possible non-linear environmental effects, as well as

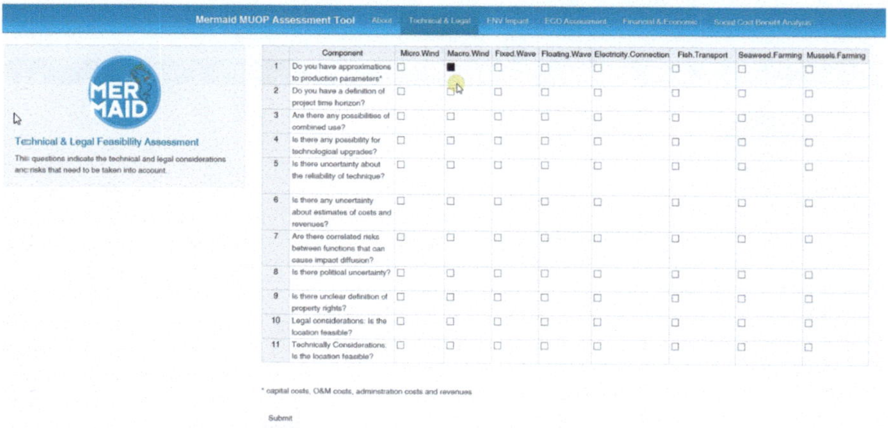

Fig. 5.1 Screenshot of the questionnaire for technical and legal feasibility assessment

the irreversible environmental effects of the operation of the platforms. The table below presents the questions posed to experts and researchers, including the set of risks to be identified. The answers of the users, which should be based on an Environmental Impact Assessment or Environmental Analysis undertaken during the design phase of the MUOP, are quantified for the tool (Fig. 5.2).

Part 3: Economic Valuation of Environmental Change In this part, the user will also have to choose the location/case study for the potential placement of the MUOP. This choice will be translated into monetary values of the relevant environmental effects, using parameter-estimated unit monetary values. CO_2 emissions' data are also inserted into the tool, annotated with category (wind, wave) and comparison method (ENTSO, COAL). CO_2 emissions' data have a single value per category and method. In this way, the corresponding change in CO_2 emissions due to MUOP operation is monetized through the social cost of carbon as an input to the social cost-benefit analysis (CBA) (Fig. 5.3).

Part 4: Financial and Economic Assessment In this part, the user will upload the data needed for the analysis, formatted as a csv (comma-separated value) file, a format that can be easily exported from all common spreadsheet software such as Microsoft Excel. This data include mainly costs and revenues for different categories (turbine, energy) and uses (wind, wave). They can also include prices (energy price). Costs, revenues, and prices are time series with a yearly resolution. Economic effects

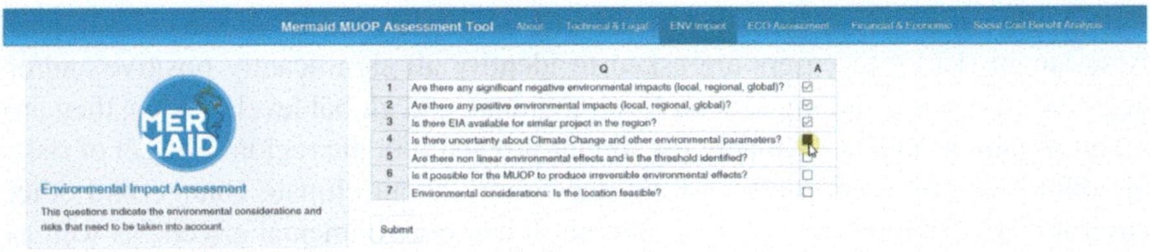

Fig. 5.2 Screenshot of the questionnaire for environmental impact assessment

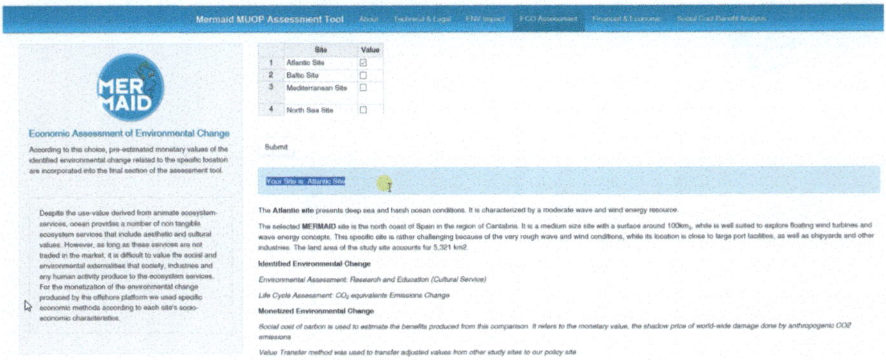

Fig. 5.3 Screenshot of site selection that allows the tool to use the corresponding predefined parameters

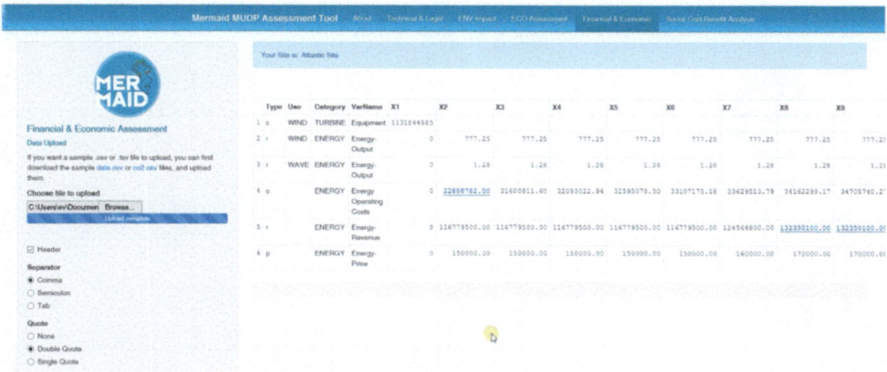

Fig. 5.4 Screenshot of data uploading for cost and benefits

concern monetized impacts of job creation and changes in the growth rate of the relevant geographical region, while unit cost is predetermined estimated values, integrated in the background data of the tool (Fig. 5.4).

Part 5: Risk Analysis The tool concludes with a risk analysis, simulating different scenarios to define the overall risk of the selected infrastructure. The tool models the input parameters as randomly distributed according to a predefined distribution (probability distributions for each variable are chosen according to the use and are predefined per site and use). With these parameters, the tool runs a Monte Carlo simulation so as to obtain a distribution for the total cost. The results are presented as a summary table with basic statistical values for the distribution of the total cost and graphic visualizations (Xepapadeas et al. 2017). It should be noted that the tool is able to compare at the same time the estimated net present value under different discount rates. Furthermore, the tool calculates and compares the net present value for the case of including the monetized externalities and for the case where these are not included (Fig. 5.5).

5.3 Common Practice Shortcomings

Oceans of Tomorrow projects are multidisciplinary projects with multiple stakeholders. This interdisciplinary nature typically results in a lack of ubiquitous language. Usually the stakeholders have noncomparable methodologies regarding data collection, data cleaning and preprocessing, data analysis, and more importantly, results interpretation. Moreover, most often socio-economic assessment is performed manually using industry standard packages that can be found in commercial products, such as Microsoft Excel. Data handling is also performed mostly manually. For example, the task of comparing two alternative scenarios may require a time-consuming process of repeating the analysis for each scenario and trying to manually compare results from multiple outputs.

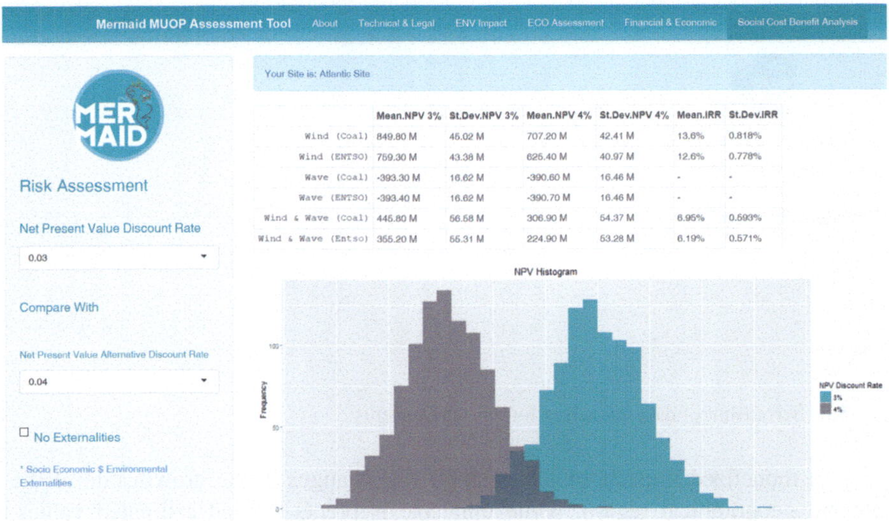

Fig. 5.5 Screenshot of results of social cost-benefit analysis-risk assessment

5.4 Detailed Exposition of Risk Analysis

The tool materialized a streamlined robust methodology for the researchers/potential decision-makers. This methodology included the data format and modeling in order to be fed into the tool, as well as the decision-making process. The tool allowed for a faster analysis, resulting in cost saving for the stakeholder. Due to the automation of risk analysis, we were able to extend data analysis abilities to a larger and more refined parameter space (Xepapadeas et al. 2017).

Monte Carlo simulations perform risk analysis by building models of possible results by substituting a probability distribution for any factor that has inherent uncertainty. It then calculates results over and over, each time using a different set of random values from the probability functions. The basic idea for all the sites/locations is the same. We create a model of the net present value (NPV), assuming a 3% and 4% discount rate (in our case) and internal rate of return (IRR), and we run Monte Carlo simulations. Monte Carlo simulations are used on stochastic models to provide various statistical information such as mean, standard deviation, skewness, and others. The input for the Monte Carlo simulations in our case is a cash flow made up of revenues and costs over a given time horizon, which is used to calculate the NPV. By modeling some or all the variables with carefully chosen probability distributions, we obtain different values of the NPV. Probability distributions are a much more realistic way of describing uncertainty in variables of a risk analysis scenario. Using Monte Carlo simulations, the NPV of a designated number of instances is stored (above 1000 instances provide appropriate accuracy in most cases), and a histogram and cumulative chart are created along with the aforementioned statistical information. When modeling random variables for our stochastic model, we consider two factors. First, we should model temporal variables, such as price, product output, etc. Second, our model distributions should be realistic; otherwise, our model may yield meaningless results. If the distributions we chose for our variables have no basis, then neither will the results of our models. For our sites, therefore, we chose to use triangular and normal distributions. For example, the triangular distribution was chosen, because in our data, we had three differing values for the same variable; hence, the triangular distribution was used to take into account all three values. In the sensitivity analysis for each site, we consider a base case for a variable that affects the cash flows, and we perturb the base case within the range of a maximum and minimum value. The sensitivity analysis calculates the net present value (NPV) for alternative scenarios for the values of this variable in steps of 10% increases from the maximum to the minimum, while the rest of the variables remain constant at their base case values. Critical variables are those with a relatively high slope in the NPV spider graph, and switching values are those in which NPV has a value of zero. Below we give a general description of the approach we used for each site.

5.4.1 *North Sea Site*

For the North Sea site, we had three platform uses that needed to be examined: energy, mussels, seaweed, and combinations of the above.

5.4.1.1 Monte Carlo Simulation

- Triangular distribution was used in mussels investment, seaweed investment, mussels operating cost, seaweed price, and seaweed operating cost.
- Normal distribution was used in energy output, mussels output, mussels price, and seaweed output.

5.4.1.2 Sensitivity Analysis

We consider the following scenarios for the purposes of sensitivity analysis. The scenarios refer to the energy, seaweed, and mussels projects.

	Min	Base[a]	Max
Mussels investment	0.7805	1.00	1.2195
Seaweed investment	0.8	1.00	1.2
Energy output	0.885	1.00	1.115
Energy operating cost	0.5919	1.00	1.4081
Mussels output	0.9375	1.00	1.0625

(continued)

	Min	Base[a]	Max
Mussels price	0.9787	1.00	1.0213
Mussels operating cost	0.261	1.00	1.739
Seaweed output	0.9625	1.00	1.0375
Seaweed price	0.5185	1.00	1.4815
Seaweed operating cost	0.812	1.00	1.188

[a]Base refers to 100% of the central value for the corresponding variable. Min and max refer to the corresponding percentages of the base case

5.4.2 *Atlantic Site*

For the Atlantic site, we had two uses that needed to be examined: wind, wave, and their combination.

5.4.2.1 Monte Carlo Simulation

- Triangular distribution was used in wind investment.
- Normal distribution was used in energy output (wind) and energy output (wave).

5.4.2.2 Sensitivity Analysis

We consider the following scenarios for the purposes of sensitivity analysis. The scenarios refer to the wind and wave project.

	Min	Base[a]	Max
Equipment	0.90	1.00	1.10
Output (wind)	0.80	1.00	1.20
Output (wave)	0.80	1.00	1.20

[a]Base refers to 100% of the central value for the corresponding variable. Min and max refer to the corresponding percentages of the base case

5.4.3 *Mediterranean*

For the Mediterranean site, the only platform use that needed to be examined was fish.

5.4.3.1 Monte Carlo

- Triangular distribution was used in wind investment, fish investment, and fish revenue.
- Normal distribution was used in fish labor, raw material, other, maintenance, operating costs, and energy output.

5.4.3.2 Sensitivity Analysis

We consider the following scenarios for the purposes of sensitivity analysis. The scenarios refer to the fish sub-project.

	Min	Base[a]	Max
Equipment (fish)	0.85	1.00	1.15
Revenue (fish)	0.75	1.00	1.25
Labor (fish)	0.75	1.00	1.25
Raw material (fish)	0.75	1.00	1.25
Other (fish)	0.75	1.00	1.25
Maintenance (fish)	0.75	1.00	1.25
Operating costs (fish)	0.75	1.00	1.25

[a]Base refers to 100% of the central value for the corresponding variable. Min and max refer to the corresponding percentages of the base case

5.4.4 Baltic Site

For the Baltic site, the only platform use that needed to be examined was energy.

5.4.4.1 Monte Carlo Simulation

- Triangular distribution was used in energy investment and maintenance.
- Normal distribution was used in energy output and artificial reefs.

5.4.4.2 Sensitivity Analysis

We consider the following scenarios for the purposes of sensitivity analysis. The scenarios refer only to the energy project. Note that due to lack of data, the NPV calculations do not include operating costs; thus, the sensitivity analysis refers to the NPV defined in terms of construction cost, maintenance cost, and revenues due to energy output and reefs. In the Monte Carlo analysis, we have calculated the maximum annual equivalent operating cost which would result in a positive NPV.

	Min	Base[a]	Max
Construction	0.8	1	1.2
Output	0.8	1	1.2
Maintenance	0.85	1	1.15
Reefs	0.75	1	1.25

[a]Base refers to 100% of the central value for the corresponding variable. Min and max refer to the corresponding percentages of the base case

5.5 Technical Characteristics

We created the decision support tool as an open-source-based web application, based on R language Shiny. R is a programming language and free software environment for statistical and graphics that is supported by the R Foundation for Statistical Computing. The R language is widely used among statisticians and data miners for developing statistical software and data analysis.[1] Shiny is an R package for building interactive web applications straight from R. Shiny Server can deliver your R visualization to your customers via a web browser and execute R functions, including database queries, in the background. The tool was built entirely with the use of open-source technologies, which allow for a low cost (both initial development cost and usage, maintenance, and update cost), are auditable, and are continuously developed and maintained from a large community.

5.6 Architecture of the Tool

The tool is designed as a typical client server web-based application. This architecture allows multiple clients (computers, tablets, smartphones) to request services from the application's server and display the results. The application runs on a Linux web server, but it can also easily be deployed in a cloud infrastructure such as Amazon Elastic Compute Cloud, allowing for a more scalable solution (Fig. 5.6).

5.7 Conclusions: Future Enhancements

Building on the experience of MERMAID's assessment tool, we propose enhancements in the following aspects:

[1]https://en.wikipedia.org/wiki/R_(programming_language)

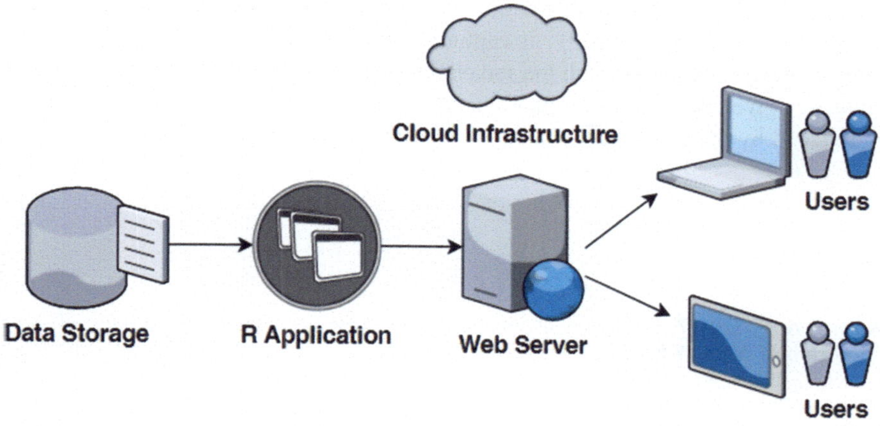

Fig. 5.6 Overview of DSS tool client – server architecture

- *Reproducibility of analysis*
  A researcher should be able to reproduce their analysis, that is, run the same model with the same assumptions based on the exact same set of data. That would allow also for comparison based on dataset enhancements and changes. For that purpose, a data versioning system could be added into the tool.
- *Automation of data preprocessing and curation*
  The tool could potentially give the user the ability to be able to preprocess raw climate data (wind, wave, currents, etc.) from existing repositories and shape them in the appropriate input format for the tool using graphical tools. That would allow for a more sophisticated model and the expansion of the tool in other domains such as spatial analysis.
- *Spatial placement of the platforms*
  The enhanced tool could include a spatial data planning component that would use relevant project data (climate, engineering, ecological, socio-economic) to present the spatial alternatives, creating a suitability map. The user should be able to revise suggested placements and evaluate new areas for placement based on the constraints fed in the tool.
- *Expandability in other projects with different socio-economic activities*
  The data modeling of the socio-economic activities can be generic enough to expand any number of characteristics.
- *Sensitivity analysis*
  The enhanced tool could visualize the outputs of sensitivity analysis as proposed in the underlying methodology. More specifically, it could allow for numerous model runs to assess how changes in model parameters will affect model outputs, thus allowing the user to assess the parameters in which the model is most sensitive to.

In the described tool could perform manually cross-validation to refine parameter selection for the simulation. In the future, this fine-tuning of the simulation could be performed automatically. MERMAID's assessment tool provided researchers with an intuitive way to evaluate multiple scenarios that would be hard and time-consuming to assess manually. It captured a lot of information carefully chosen after multiple interactions with the stakeholders of the project MERMAID, a unique challenge given the multidisciplinary nature of the project, as well as the complexity introduced by the concurrent evaluation of different geographical sites.

References

Koundouri, P. (2017). The ocean of tomorrow, investment assessment of multi-use offshore platforms: Methodology and applications – Volume 1. Springer International Publishing. eBook ISBN: 978-3-319-55772-4, Hardcover ISBN: 978-3-319-55770-0. https://doi.org/10.1007/978-3-319-55772-4.

MERMAID. (2014). FP7 European Commission project. Online platform. www.mermaidproject.eu

Xepapadeas, P., Giannouli, A., Koundouri, P., Moussoulides, A., Tsani, S., & Xepapadeas, A. (2017). Risk analysis for the selected MERMAID final designs. In P. Koundouri (Ed.), *The ocean of tomorrow. Environment & policy* (Vol. 56). Cham: Springer.

Chapter 6
Techno- and Socio-economic Models of Production with Application to Aquaculture: Results from the BlueBRIDGE Project

Gerasimos Antzoulatos, Charalampos Dimitrakopoulos, Eleni Petra, Stella Tsani, and Phoebe Koundouri

Abstract This chapter (This work has received funding from the European Union's Horizon 2020 research and innovation program under grant agreement N° 675,680) presents in brief the work completed within the scope of the BlueBRIDGE project with the aim to support the sustainable management of aquaculture production. The work completed is the result of an interdisciplinary team of IT experts, economists, biologists, social scientists, and marine professionals that have joined forces with the aim to develop IT advanced tools and integrated models of production. A very important part of work within the scope of BlueBRIDGE concerned Cloud Computing Infrastructure, Virtual Research Environments (VRES), and the development of integrated production models with focus on aquaculture. These can support well-informed site management and sector-related decision-making.

G. Antzoulatos
Oloklirомena Pliroforiaka Sistimata, Athens, Greece

C. Dimitrakopoulos
Communication & Information Technologies Experts, Kesariani, Greece

E. Petra (✉)
National and Kapodistrian University of Athens, Athens, Greece
e-mail: elenipetra4@gmail.com

S. Tsani
University of Ioannina, Ioannina, Greece

International Centre for Research on the Environment and the Economy, Athens, Greece

P. Koundouri
School of Economics and ReSEES Laboratory, Athens University of Economics and Business, Athens, Greece

United Nations Sustainable Development Solutions Network Europe, Paris, France

EIT Climate–KIC Hub Greece, ATHENA Research and Innovation Center, Athens, Greece

ICRE8: International Centre for Research on the Environment and the Economy, Athens, Greece

P. Koundouri (ed.), *The Ocean of Tomorrow*, Environment & Policy 57,
https://doi.org/10.1007/978-3-030-56847-4_6

Keywords Virtual Research Environments · Cloud Computing Infrastructure · Aquaculture · BlueBRIDGE

6.1 Introduction

BlueBRIDGE, a HORIZON 2020 project responded to the Juncker Investment Plan opening funding opportunities for e-infrastructures, innovating current practices in producing and delivering scientific knowledge advice to competent authorities.[1] The project aimed at the enlargement of the spectrum of growth opportunities in distinctive Blue Growth areas. BlueBRIDGE has built on existing EU and international e-infrastructures providing capacity-building in interdisciplinary research communities of scientists, data managers, and educators in academic institutions and industries focusing on four major challenges: (1) stock assessment, (2) socio-economic performance analysis in aquaculture, (3) fisheries and habitat degradation monitoring, and (4) education and knowledge bridging on protection and management of marine resources. BlueBRIDGE capitalized on past investments and used a proven e-infrastructure connecting 1500+ scientists, integrating +50 repositories, executing +13,000 models and algorithms/month, and providing access to over a billion quality records in repositories worldwide, with 99.7% service availability. BlueBRIDGE also focused on user needs, opening services and data to actors and liaising with competent agencies and SME Innovation Clusters.

BlueBRIDGE brought together an authoritative and complementary consortium with expertise in multiple scientific domains. It bundled forces from International Government Organizations (FAO, Italy; ICES, Denmark; ASSOCIATION POLE MER BRETAGNE (PMBret), France), research institutes (CNR, Italy; ERCIM, France, IRD, France; FORTH, Greece), STIFTELSEN GRID ARENDAL (GRID Arendal, Norway), industry ENGINEERING – Italy, CLS-France), and SMEs (TRUST-IT, United Kingdom, OLOKLIROMENA PLIROFORIAKA SISTIMATA (I2S), Greece, COMMUNICATION & INFORMATION T ECHNOLOGIES EXPERTS ANONYMOS ETAIREIA SYMVOULEFTIKON KAI ANAPTYXIA KON YPIRESION (CITE), Greece), education and computer science domains (ETHNIKO KAI KAPODISTRIAKO P ANEPISTIMIO ATHINON (UOA), Greece), establishing a network with a proven track in VREs and e-infrastructures, marine, environmental and fisheries, and science and economy.

A very important part of work within the scope of BlueBRIDGE concerned Cloud Computing Infrastructure, Virtual Research Environments (VRES), and the development of integrated production models with focus on aquaculture. The VRES aimed to deliver economic growth and environmental planning capacity-building instruments to confront modern science and economic analysis challenges, as identified by both scientists and aqua-farming industry and investors. VRES have been

[1] See: https://www.bluebridge-vres.eu/.

developed within the project with the aim to allow for the first time to address in an integrated way the problem of global knowledge access, seamless data exchanges, and reuse between aquaculture companies and its related stakeholders. The tools have been developed through the cooperation of different experts (computer scientists, biologists, economists, sociologists, etc.) with the intention to give new instruments to policy makers and scientists for supporting their research and well-informed decisions.

The technology developed allows (a) companies to improve efficiency, increase profitability, and do business in a sustainable, environmentally friendly way, enabling them to convert their operation data into knowledge and actionable results; (b) researchers access datasets concerning the practices and performance of aquaculture producers, generating new knowledge and evaluating the practical indicators of aqua-farming performance; (c) governments and environmental agencies to evaluate the current situation and define policies; (d) SMEs and academic institutes to provide training for executives and decision-makers on the sector; and (e) investors to efficiently take decisions using, for instance, multifactor evaluation approaches. The delivering of these technologies has been completed through a continuous development process in which (i) detailed needs for related capacity-building instruments are collected and verified, (ii) service and data offerings are delivered, and (c) results and effectiveness of the instruments are validated.

Integrated production models developed within the scope of the project try to fulfill the basic demand of incorporating all the possible factors that are vital for global sustainable development within the process of productivity-potential. The methodology proposed and the type of models that have been developed are easily applicable to several different domains such as agriculture, food production, industry, etc. Integrated production models are developed for the aquaculture sector, in order to (1) assess and conceptualize the social, economic, and environmental impacts of climate change, over the same time period; (2) gather intelligence and data that will assist decision-making plagued with uncertainty, (3) foster innovation, and (4) strengthen the transition towards a Blue Economy by incorporating specific socio- and techno-economic factors. Sustainability indicators within a socio-, techno-, and ecological context are utilized in order to examine and record developments over a specific time period and assist in producing qualitative and quantitative data, within a much larger system that includes consumers and the chain of production.

This chapter provides a brief overview of the VREs developed within the BlueBRIDGE project with focus on aquaculture. Section 6.2 discusses VRES for aquaculture. Section 6.3 reviews the socio-economic extensions of the production models developed. Section 6.4 briefly presents the option to run and examine alternative future production scenarios. Last section concludes.

6.2 VRES for Aquaculture Production

Two identified VREs have been examined within the BlueBRIDGE scope. The first includes performance evaluation, benchmarking, and decision-making in aquaculture VRE. This approach focuses on capturing and confronting the dual challenge of understanding (a) the performance of an aqua-farming operation, allowing on one hand investors and entrepreneurs to conform to environmental rules and optimize the use of resources (monetary and nonmonetary ones), and (b) the pressure on investment and on environment produced by such operations, so that scientists and policy makers can craft guidelines or even regulations, taking into account the economic interest and socio-economic impact of those operations. To achieve the goal, a VRE is provided, granting aqua-farming sector entrepreneurs, investors, scientists, and decision-makers with access to (a) open data combining performance, environment, regional, and socio-economic datasets, (b) a framework to support performance analysis and benchmarking, and (c) a fundamental baseline set of indicators and models that provide the performance metrics in question.

In order to accomplish the goals, the provided services consist of three main components, namely, site management, model management, and What-If Analysis. In site management, aqua farmers set up the profiles of their sites (thermal and environmental characteristics) to provide useful information to the model's creation. In model management, the supplied historical production data in combination with environmental and thermal site profile are analyzed to create models which are capable to evaluate the production performance. The What-If Analysis service (discussed in more detail in Sect. 6.4 of this chapter) provides to fish farm managers the opportunity to determine what-if scenarios, evaluate the vital key performance indicators for the fish growth, and make efficient and accurate production plans. The service is developed with the intention to allow for future benchmarking analysis, comparing one's company key performance indicators (KPIs) against other aquacultures which operate under "similar" circumstances.

The second VREs option examined within the BlueBRIDGE scope regards Strategic Investment Analysis and Scientific Planning and Alerting. This alternative aims to satisfy the need for intelligent identification of locations of interest under an open-ended set of criteria, as required by both investors seeking optimization of intended investments and by scientists seeking areas that are becoming of environmental importance. Building on the products of the previous as other VRE offerings (services, models, and datasets) and extending those with techniques of computational intelligence, more algorithms/indicators/datasets, this VRE delivers a cutting-edge geospatial multifactor optimization and alerting platform of unprecedented value tool for scientists and the aqua-farming industry and vast opportunities for interdisciplinary exploitation beyond its definition.

This VRE is based on a cost-driven techno-economic evaluation model for aquaculture investment analysis. Breaking down the costs of building and running an aqua farm while estimating sales from its' main business activities, the model can provide a 10-year estimation for the cash flows of the specific aqua farm. This model also takes into consideration the output of production models in order to estimate costs of production, and it is currently being enhanced with variables of socio-economic impact. Net present value (NPV), internal rate of return (IRR), yearly net profit margin, and several other KPIs and financial metrics are produced from this model aiming at a more complete investment analysis that will support a solid scientific planning and decision-making procedure for aquaculture.

6.3 Socio-economic and Environmental Monetization Extension in Aquaculture Management Tools

A core advancement within the scope of the BlueBRIDGE project is the integration of social, economic, and environmental variables in the decision support tools developed for aquaculture production. The research on the modeling of aquaculture production is ongoing. Available models look at the financial and economic aspects of aquaculture, while others are developed along the lines of cost-driven production. The work completed for BlueBRIDGE brings socio-economic and environmental impact into consideration when analyzing aquaculture operation performance. The multi- and intradisciplinary work completed for the project responds to challenges of identifying, conceptualizing, and monetizing the social and environmental impact of aquaculture. Once identified these model parameters have been combined in a workable way with specific techno-economic and production models of Blue Economy, considering data and computational resources in reach.

For this the research team has undertaken a thorough social cost benefit analysis (SCBA) in which the costs and benefits of fish production are identified, modeled, evaluated, and monetized. SCBA aims at incorporating the effects that extend beyond private costs and revenues and regard the broader impact which aquaculture has on the society. The SCBA systematically identifies, organizes, and valuates the benefits and the costs of aquaculture. The approach is detailed in Tsani and Koundouri (2018) and it aims at (i) extending the identification of the costs and benefits of aquaculture beyond pure financial metrics and monetary terms and (ii) extending the modeling of aquaculture production beyond the quantification of private costs and benefits and including in the analysis the social costs and benefits associated with aquaculture (a graphical summary of the methodology is provided in Fig. 6.1). The modeling approach consists of three parts: (i) ecological part where the modeled relationships aim at capturing the interactions of aquaculture and associated costs and benefits as regards the environment (CO_2 emissions, water pollution, spatial limitations etc.); (ii) economic part which makes provision for the explicit incorporations of economic determinants in production models; and (iii) social part that introduces social costs and benefits in the aquaculture production and management decision process. Appropriate relationships are formulated which quantify and introduce the socio-economic and environmental costs and benefits of aquaculture into the decision support system of aquaculture management.

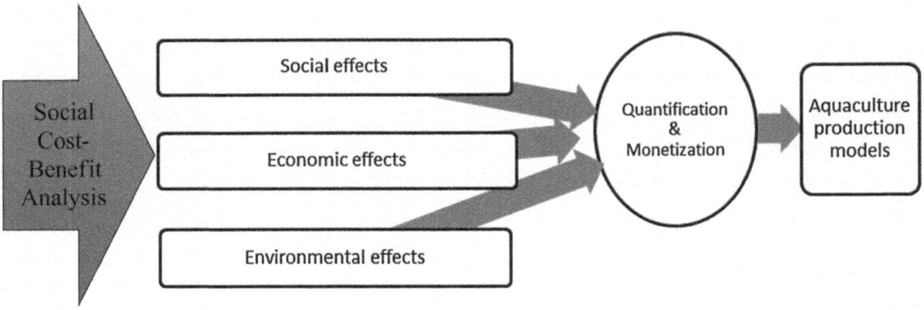

Fig. 6.1 Identification of socio-economic and environmental impact of aquaculture and their integration to aquaculture production models. (Source: Tsani and Koundouri 2018)

The costs and benefits considered include among other financial production, social interactions, and environmental impacts, translated into effects on ecosystem services and social effects evaluated over time and space. This framework allows decisions that are consistent with the concepts of environmental sustainability, economic efficiency, and social equity. The main goal of this work has been the contribution to an interdisciplinary and participatory framework of analysis of technical, environmental, economic, social aspects of Blue Growth, which can provide policy recommendations for improved implementation of the Initiative, allied with the relevant Sustainable Development Goals (SDGs) and the Marine Strategy Framework Directive (MSFD) and Marine Spatial Planning (MSP).

Drawing on the resent research, the working team on this model extension has identified and quantified the socio-economic and environmental costs and benefits associated to aquaculture in a way compatible to the techno-economic and cost-driven production models available. The work undertaken allows for the development of an integrated production model for aquaculture that takes into consideration and incorporates the complex feedbacks between ecological and economic aspects of aquaculture production. In this way (i) it can be met the integrated management of aquaculture production that takes into consideration both the private costs and benefits but also the social costs and benefits associated with externalities and effects not appropriately captured by market-driven functions and factors, and (ii) it can be provided quantified insights to the social costs and benefits that producers internalize or can internalize, which can complement policies targeting aquaculture management and financing (e.g., subsidies, environmental taxes, etc.).

6.4 What-If Analysis in Aquaculture Production Models

In aquaculture, the accurate estimation of the fish growth is a cornerstone of an efficient and profitable production. Without information on expected growth, it would be impossible to assess feed requirements for specific period and additionally to approximate the date that fish are ready for harvest. Growth depends upon the quantity and quality of food ingested, as well as environmental factors such as sea temperature, oxygen, currents, etc. In recent decades, the prediction of growth rate and feed consumption of cultivated fish have gained increasing attention, and many empirical models based on mathematical/statistical methods have been proposed. Models can be used to estimate the growth and feed intake in terms of specific factors rely on the experimental data and the experts' knowledge (Petridis and Rogdakis 1996; Zhou et al. 2017; Jobling 2008; Mayer et al. 2008; Arnason et al. 2009).

A core output of the work within the BlueBRIDGE project, has been the creation of models able to predict production indicators such as feed conversion rate (biological FCR), Feeding Rate (FR), and mortality rate based on the study of historical production data. The models are established via the state-of-the-art regression methods, such as generalized additive models (GAMs) and multivariate adaptive regression splines (MARS). For each indicator the model with the highest accuracy is employed in order to estimate the values of indicators in various growth levels (fish weight) and different temperatures. The outcome of this process is a representation of empirical models, which simulates the relationship between growth, feeding, and temperature. Specifically, the modeling results in the development of tables for biological FCR, feeding rate and mortality rate in terms of fish weight and temperatures. Figure 6.2 provides an example graphical representation of a biological FCR table indicative of the developed model. For each weight category and temperature, it can be estimated the indicator value by the modeling of real production data.

Having these information aqua farmers can perform production plans and assess vital key performance indicators (KPIs) by the examination different hypothetical statements (what-if scenario). Specifically, they can define easily the conditions of hypothetical scenarios by setting the number and the average weight of fish population and the period of interest that cultivate the fish (stocking date and harvest date) and forecast the evolution of KPIs in this period. Through this simulation process, they can alter the hypothesis and compare the results of each scenario in terms of

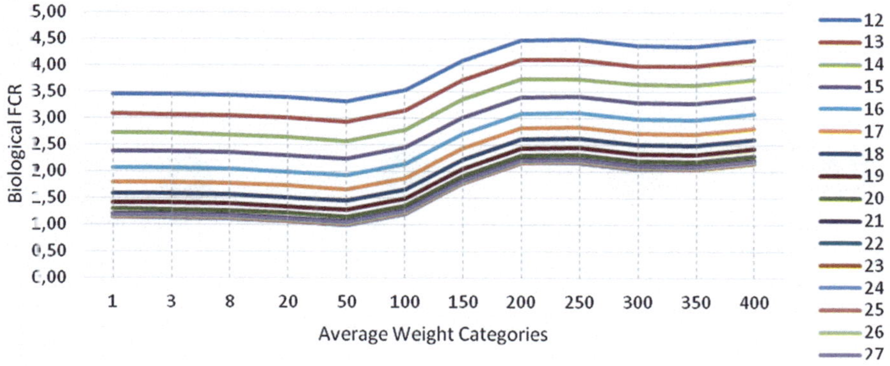

Fig. 6.2 Representation of biological FCR table per weight category grouped by temperature

LTD Biological FCR, LTD Economical FCR, LTD SGR, LTD Growth, and LTD Mortality. Simultaneously, they can benchmark their aqua farm performance against the competition. The core idea behind benchmarking process is to provide a tool in which aqua farmers can compare their farms' performance against the performance of other farms with similar characteristics in terms of the crucial KPIs without sacrificing the confidentiality and privacy of their data. More specifically, using Blue Economy tools, aqua farm managers achieve to estimate and benchmark the performance of their farms' production by carrying out simple steps as illustrated in Fig. 6.3 and discussed in brief next.

Setup the site of interest (Setup Site tool): The user can define the specific environmental characteristics of the site of interest providing the average temperature fortnightly and the geographical location of the site.

Create models (Setup Model tool): The user can develop reliable and powerful machine learning models, which are capable to estimate vital production indicators, such as biological FCR, FR, and mortality rate, providing historical data and details regarding the production of the specific fish species of the site of interest.

Create a hypothetical scenario (What-If Analysis tool): The user can draw a hypothesis and evaluate it by using an already existing model. The system using the user-defined conditions and the output of the modeling process (tables of FCR, FR, and mortality rate) can grow the population up to a particular date (harvest date).

Evaluate and benchmark the performance: The results of the previous step are presented by various interactive graphs and tables exhibiting the performance of key performance indicators (LTD Biological FCR, LTD Economical FCR, LTD SGR, LTD Growth, LTD Mortality, monthly feed consumption, and average weight per day). In addition, system compares the performance of user's production against the competition. The benchmarking process is carried out over the sites with similar environmental characteristics implying that they will have similar productions. The system executes a back-end process hidden from the end users and detects all the sites, which are similar with the site of interest. Then, it creates a "global" production model. The same what-if scenario is fed to the global model, and the results are compared with the results coming from aqua farm's production model. The process is iterative and users can go backward to create new sites and/or models or evaluate new scenarios and so on.

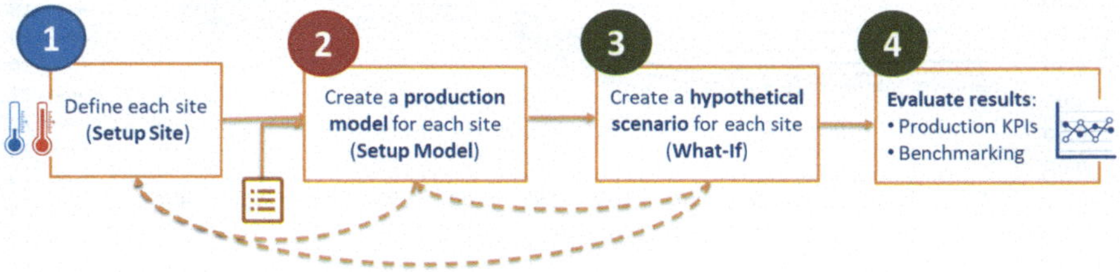

Fig. 6.3 Steps of performance evaluation and benchmarking process

6.5 Concluding Remarks

Aquaculture can be core to Blue Growth targets with benefits exceeding the private benefits. The sustainable and efficient management of aquaculture production, at micro- but also at macro- level, requires the use of cutting-edge technology, novel IT applications, and integrated socio-economic tools. The tools and the methodology developed within the scope of the BlueBRIDGE project allow for the use of advanced IT applications and facilities and the introduction of the wider socio-economic and environmental effects of aquaculture into the production management. This can support well-informed management and decision-making. The tools and methods developed facilitate the estimation of an integrated value of production that looks beyond output maximization. Developing and proposing easy-to-use tools enable all producers and the sector overall to engage in the technology race and use it at their own benefit. From a policy perspective, the outputs of the project enable the well-informed and forward-looking decision-making and target setting.

References

Arnason, T., Bjornsson, B., Steinarsson, A., & Oddgeirsson, M. (2009). Effects of temperature and body weight on growth rate and feed conversion ratio in turbot (Scophthalmus maximus). *Aquaculture, 295*, 218–225.

Jobling, M. (2008). Environmental factors and rates of development and growth. In P. J. Hart & J. D. Reynolds (Eds.), *Handbook of fish biology and fisheries* (Volume 1: Fish biology) (pp. 97–122). Blackwell Publishing Ltd. https://doi.org/10.1002/9780470693803.ch5.

Mayer, P., Estruch, V., Blasco, J., & Jover, M. (2008). Predicting the growth of gilthead sea bream (Sparus aurata L.) farmed in marine cages under production conditions using temperature-and time-dependent models. *Aquaculture Research*, 1–7.

Petridis, D., & Rogdakis, L. (1996). The development of growth and feeding equation for sea bream, Sparus aurata L., culture. *Aquaculture Research, 27*, 413–419.

Tsani, S., & Koundouri, P. (2018). A methodological note for the development of integrated aquaculture production models, with P. Koundouri. *Frontiers in Marine Science, 4*, 406. https://doi.org/10.3389/fmars.2017.00406.

Zhou, C., Xu, D., Lin, K., Sun, C., & Yang, X. (2017). Intelligent feeding control methods in aquaculture with an emphasis on fish: A review. *Reviews in Aquaculture*, 1–19.

Chapter 7
Increasing Land-Sea Synergies and Coastal-Rural Collaboration for a Healthy Ocean: The COASTAL Project

Ebun Akinsete, Alice Guittard, and Phoebe Koundouri

Abstract Coastal and sea regions not only concentrate populations and intensive economic activity but also environmental stresses and higher levels of pollution. On the other hand, rural hinterlands face depopulation and often economic recession while still environmentally impacting coastal regions. Land-based activities (agriculture, forestry, industries and urbanization) are directly and indirectly impacting land, coastal and sea ecosystems (soil and river pollutions, marine water eutrophication, etc.), and the coastal regions are the downstream recipient of these land use negative practice externalities. The EU H2020 COASTAL project, presented in this chapter (This work has received funding from the European Union's Horizon 2020 research and innovation program under grant agreement N° 773,782.), adopts an innovative source-to-sea approach where at river basin scale, sea, coastal and rural regions are seen as a whole and single ecosystem. The project seeks to improve land-sea synergies in strategic business and policy decision making, and collaborations between coastal and rural stakeholders in order to (1) create connections between

E. Akinsete (✉)
ICRE8: International Centre for Research on the Environment and the Economy, Athens, Greece

United Nations Sustainable Development Solutions Network, Athens, Greece

EIT Climate-KIC Hub, Athens, Greece
e-mail: ebun.akinsete@icre8.eu

A. Guittard
ICRE8: International Centre for Research on the Environment and the Economy, Athens, Greece

United Nations Sustainable Development Solutions Network, Athens, Greece

P. Koundouri
School of Economics and ReSEES Laboratory, Athens University of Economics and Business, Athens, Greece

United Nations Sustainable Development Solutions Network Europe, Paris, France

EIT Climate–KIC Hub Greece, ATHENA Research and Innovation Center, Athens, Greece

ICRE8: International Centre for Research on the Environment and the Economy, Athens, Greece

P. Koundouri (ed.), *The Ocean of Tomorrow*, Environment & Policy 57,
https://doi.org/10.1007/978-3-030-56847-4_7

land ecosystems and coastal ecosystems, for sustainable use and management of theses ecosystems as well as reaching the Sustainable Development Goals (https://sustainabledevelopment.un.org/?menu=1300) (focusing on SDG 6, 13, 14, 15); (2) support sustainable growth in rural, coastal and marine sectors by fostering cross-sectoral collaborations; and (3) develop links between EU WFD, MFD and CAP by developing policy alternatives for coastal-rural areas. The COASTAL project adopts a strong iterative participatory multi-actor, bottom-up approach based on collaboration between rural, coastal and sea stakeholders via Multi-Actor Labs (MALs) centred around six selected coastal regions (in Belgium, Sweden, Romania, Greece, Spain and France) with both common and unique opportunities and challenges. The methodology and tools used combine local and scientific knowledge to explore and analyse social, environmental and economic land-sea interactions in a collaborative System Dynamics framework to identify problems and develop practical and robust business road maps and strategic policy guidelines to improve land-sea synergies and coastal-rural collaborations.

Keywords COASTAL · Source-to-sea approach · SDGs · Rural sustainable growth · Multi-actor · Bottom-up approach

7.1 The Need to Increase Land-Sea Synergies

7.1.1 Coastal Areas and Rural Hinterland

Nearly 40% of the world's population lives within 100 km of the coast. Coastal areas experience higher population density, higher population growth associated with global trend of coastal migration and higher urbanization (most of the world's megacities are located in the coastal zone cf. Brown et al. 2013) than the hinterland (Small and Nicholls 2003; Smith 2011; McGranahan et al. 2007). Coastal areas are also economically highly dynamic, with a high concentration of activities (on land and at sea) and a generally lower unemployment rate (in half of EU coastal regions, unemployment is lower than the national average[1]). Coastal zones are generally driven by a strong tourism industry in constant growth (in regions such as the Mediterranean, tourism activities represent 11% of regional GDP and 11% of total employment[2]). In addition to tourism, coastal areas also benefit from increased maritime activity (shipping, fishing, mining, offshore renewable energies, aquaculture). In contrast, the global phenomena of industrialization and urbanization create

[1]Eurostat regional yearbook 2011 – Coastal region, https://ec.europa.eu/eurostat/documents/3217494/5728589/KS-HA-11-001-13-EN.PDF/c0dd33ed-0db2-4d8b-ae03-26d9bf3e57fc?version=1.0

[2]Plan Bleu paper n°17, Tourism, May 2017, http://planbleu.org/sites/default/files/publications/cahier17_tourisme_en_web.pdf

lasting impact on rural hinterlands which suffer from depopulation and economic decline (Li et al. 2019). 92% of the EU territory is considered to be rural,[3] while only 28% of the EU population lives in this area.[4] The main rural sectors (agriculture, forestry and other nature-based industries) have constantly increased their capital to labour ratio, leading to a decrease in employment rates, thereby making urban and coastal areas more attractive from an economic standpoint, and pushing rural populations to migrate to these areas. The constant population growth and higher concentration of economic activity in the coastal zones give rise to an intensification of anthropogenic pressures on the rich but fragile coastal and marine ecosystems. Moreover, poor management and use of the natural resources inland further increases the environmental stresses over these ecosystems.

7.1.2 *Inland Impacts on Coastal and Sea Regions*

The source-to-sea continuum considers the land and the sea as a single component, i.e. from land to fresh water, delta, estuarine, coastline, near shore and ocean, connected through flows of waters, including sediment, pollutants, materials, biota and related ecosystem services (Granit et al. 2017). The sea, the coastal area and its hinterland form a single, unique ecosystem. The sea and coastal areas being the natural continuity of inland areas, land-based ecosystems and sea-based ecosystems are intrinsically linked, with one not only benefiting from the other but also impacting it and vice versa. Today, marine and coastal resources are increasingly jeopardized by upstream activities on land and along rivers.

Inside a river basin, land-based activities (agriculture, forestry, heavy industry and urbanization) inevitably impact coastal areas and marine water due to unsustainable land use, soil degradation and pollution of inland fresh water. As such, coastal areas are particularly vulnerable as they are subject to both land-based and marine-based human pressures (Halpern et al. 2008). Land-sourced pollution (high levels of nutrients loads) related to poor agricultural practices and continual urban development (i.e. inadequate urban sewage system) lead to marine ecosystem degradation, harmful algae blooms and eutrophication in coastal water (Howarth et al. 2002; DeGeorges et al. 2010; Le Moal et al. 2019). Waste production inland also has a severe effect on the coast and the sea, in the form of increased marine litter – approximately 8 million tons of plastic enter the ocean from land-based sources annually,[5] and constitutes more than 80% of marine litter; modelling the drift patterns of floating marine litter shows that the plastic pollution of almost every

[3]EU strategic guidelines for rural development, https://eur-lex.europa.eu/legal-content/FR/TXT/PDF/?uri=CELEX:32006D0144&from=EN

[4]Statistics on rural areas in the EU, 2017 https://ec.europa.eu/eurostat/statistics-explained/index.php/Statistics_on_rural_areas_in_the_EU

[5]SIWI Policy Brief 'Transboundary waters: cooperation from source to sea'

country is mainly caused by its own terrestrial sources (Bigagli et al. 2019). Hydraulic infrastructure has resulted in over half of the world's major rivers being severely affected by the alteration and fragmentation of their flow regimes (Nilsson et al. 2005), with deltas increasingly vulnerable to flooding and submergence as a result of the combined effects of the trapping of sediment behind dams and sea-level rise due to climate change (Syvitski et al. 2009) and in some cases over-abstraction of groundwater (Erban et al. 2014). In other cases, river flows hardly reach the sea, thus creating disequilibrium in water, sediment and nutrient supply to coastal ecosystems. Additionally, climate change, and primarily sea-level rise, poses a great threat on land to many coastal regions with populations, economic activities and coastal ecosystems put at risk.

7.1.3 The Source-to-Sea Approach Supported by a Global Policy Framework but Locally the Governance Still Fragmented

Given the scale of the challenge, the international community has recognized the need for integrated coastal and freshwater management. The UN Environment Programme Global Programme of Action for the Protection of the Marine Environment from Land-based Activities[6] (GPA) was adopted in 1995. This was followed by the Manila convention in 2012, which committed 'to improve cooperation and coordination at all levels to deal with issues related to oceans, coasts, islands and their associated watersheds, by applying integrated management such as "ridge to reef" approaches, involving stakeholders and developing innovative solutions to improve or resolve identified problems'. The importance of improving fresh and marine water quality, as well as land-based ecosystem, is also recognized by several goals within the Agenda 2030 for Sustainable Development (SDGs 6, 14 and 15).[7]

However, in many cases, regional and national policies fail to align with the international agenda. At EU level, the Maritime Spatial Planning Directive (MSPD) refers explicitly to the need to account for land-sea interactions for addressing nutrient loads; however, the quality and ecosystem status of coastal waters, regulated by the EU Water Framework Directive (WFD), is not yet adequately integrated into the Marine Strategic Framework Directive (MSFD) and the MSPD (Borja et al. 2010) (Fig. 7.1).

---

[6]'The Global Programme of Action is designed to be a source of conceptual and practical guidance to be drawn upon by national and/or regional authorities for devising and implementing sustained action to prevent, reduce, control and/or eliminate marine degradation from land-based activities'. cf. https://sustainabledevelopment.un.org/partnership/?p=7432

[7]https://sustainabledevelopment.un.org/?menu=1300

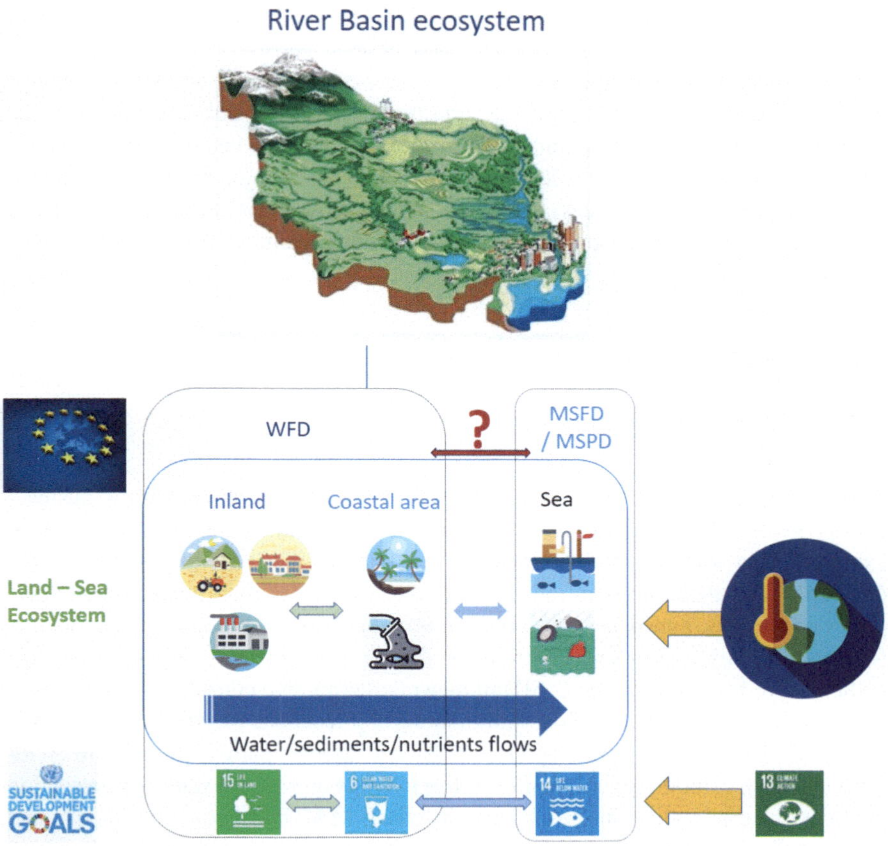

Fig. 7.1 Alice Guittard, Mapping EU and international policy frameworks onto the source-to-sea continuum. (Icones from Freepik in Flaticon.com)

7.1.4 Promoting Land-Sea Synergies to Foster the 'Source-to-Sea' Management Approach

Bringing together stakeholders from land, coastal and sea regions to create synergies and elaborate innovative approaches to solve common environmental issues in a collaborative way is necessary. Increasing land-sea synergies by creating new business opportunities in the context of the Blue and Green Growth will enable the sustainable development of sea, coastal and rural regions, making the whole land-sea ecosystem more resilient. This requires a paradigm shift at policy level, towards the adoption of source-to-sea concept, in order to ensure that policies such as the MSFD (Marine Strategy Framework Directive), WFD (Water Framework Directive) and the CAP (Common Agricultural Policy) are well connected and truly represent the Integrated Coastal Zone Management approach as well as supporting rural development and Blue Growth potentials.

Working within this context, the EU H2020 COASTAL project seeks to improve land-sea synergies in strategic business and policy decision-making, by fostering collaborations between coastal and rural stakeholders in order to support sustainable growth in rural, coastal and marine sectors by cross-sectoral collaborations and the codesign of robust land-sea sustainable strategies; create links between EU WFD and MSFD by developing policy alternatives for sea-coastal-rural areas; highlight the connections between land ecosystems and coastal ecosystems, for a sustainable use and management of these ecosystems; and reach the sustainable development goals.

7.2 The H2020 COASTAL Project

7.2.1 Methodology and Tools

The COASTAL project is a unique collaboration of coastal and rural business entrepreneurs, administrations and stakeholders, along with natural and social science experts. Local and scientific knowledge are combined to identify problems and develop practical solutions in the form of robust business road maps and strategic policy guidelines, aimed at improving land-sea synergy. A multi-actor approach is followed to analyse the social, environmental and economic land-sea interactions in a collaborative System Dynamics (SD) framework, taking into consideration the short-, mid- and long-term impacts of decision-making and feedback mechanisms on coastal and rural development. The project is organized around participatory Multi-Actor Labs (MALs), combining tools and expertise to focus on six case studies representing the major coastal regions in the EU territory. Each MAL consists of local actors and experts who participate in collaborative exercises to analyse problems and causes, propose and discuss solutions as well as validate and interpret the impacts of simulated business and policy decisions. The MALs are connected via an online platform for collaborative knowledge exchange – the COASTAL platform. The COASTAL platform and the synergistic tool sets will be further exploited and developed beyond the project lifetime. The ultimate ambitions of COASTAL are to inspire strategic land-sea planning and contribute to the formulation of integrated coastal-rural regulations at the regional, national and EU level (Fig. 7.2).

7.2.2 COASTAL Project Case Studies[8]

The COASTAL project's six case studies and corresponding MALs differ by problem context, spatial scale and social-environmental conditions. They are

[8]All case study descriptions have been developed by local MAL leaders as part of the COASTAL project and are extracted from the project internal documentations (Belgium, Jean-Luc de Kok

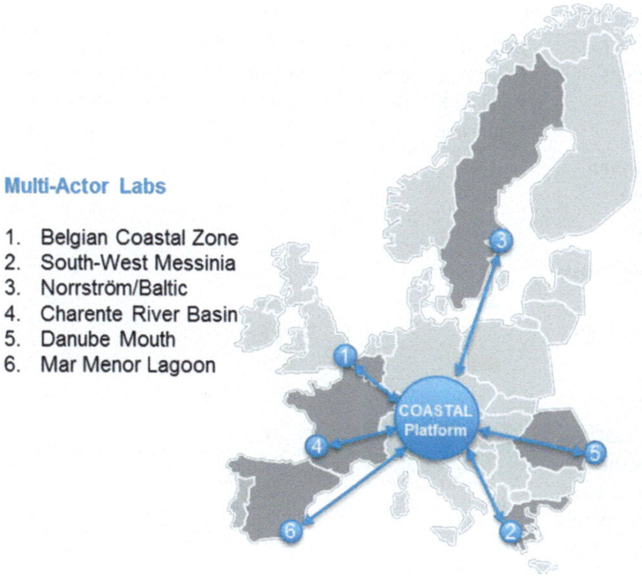

Fig. 7.2 Location of the six COASTAL case studies (cf. COASTAL project)

collectively representative of the diversity of European coastal regions with common issues and opportunities as well as local specificities (see Table 7.1).

7.2.2.1 Belgian Coastal Zone (North Sea Region)

The Belgian coast (60 km length) and hinterland face environmental and economic stresses from intensive multifunctional use of space. Land- and sea-based activities such as agriculture, fisheries, agro-food industry, transport, energy production and recreation are closely interwoven and compete for space. New development opportunities for this densely populated region are created by Blue Growth, in particular energy production (both on- and offshore) which creates opportunities for new jobs and strategic specialization of port activities – this includes innovative production methods using wave and tidal energy. Belgium is world leader in terms of know-how related to offshore energy production and the first country to implement the multipurpose use of offshore wind farms (i.e. combined with shellfish aquaculture). That said, the quality of freshwater resources is under pressure, and land-based

(VITO), Bastiaan Notebaert (VITO); France, Françoise Vernier (IRSTEA), Jean-Marie Lescot (IRSTEA); Romania, Luminita Lazar (NIMRD), Ruxandra Pop (ICEADR); Sweden, Georgia Destouni (SU), Samaneh Seifollahi-Aghmiuni (SU), Zahra Kalantari (SU); Greece, Giorgos Maneas (SU), Aris Karageorgis, (HCMR), Håkon Berg (SU); Spain, Javier Martínez-López (CSIC), Joris de Vente (CSIC))

Table 7.1 Issues highlighted in COASTAL local sectoral workshops

Issues identified for COASTAL case's study	Belgian Coastal Zone and Hinterland	Charente Basin (France)	Mar Menor (Spain)	SW Messinia (Greece)	Danube Mouth and Black Sea (Romania)	Norrstrom and Baltic (Sweden)
Water quality (and eutrophication)	✓	✓	✓	✓	✓	✓
Water quantity	✓	✓	✓	✓	✓	✓
Flood Risk and Coastal defence	✓		✓		✓	✓
Soil quality (and soil's salinization)	✓		✓	✓	✓	✓
Beach erosion				✓	✓	
Stakeholders conflicts / lack of cooperation	✓	✓	✓	✓	✓	✓
Lack of Information / Education regarding environmental issues and policies			✓	✓	✓	✓
Public awareness and lifestyle (including food habits)	✓		✓	✓	✓	✓
Waste management (inland, beach and marine litters)	✓			✓	✓	✓
Biodiversity loss	✓		✓	✓	✓	✓
Natural protected area and other Policy and management related issues management	✓		✓	✓	✓	✓

(continued)

Table 7.1 (continued)

Nature conservation	●		●	●	●	●
Cultural conservation			●	●	●	●
Traffic congestion / transport network issues	●		●	●		●
Land price/ land availability / increase urbanisation	●	●	●	●	●	●
Climate change	●	●	●	●	●	●
Sustainable growth			●	●	●	●
Seasonal population variability	●		●	●	●	●
Social challenges					●	●
Historic legacy sources of nutrients and pollutants						●
Lack of infrastructures for further development				●	●	
Taxation issue				●	●	
the need of improved renewable energy	●			●	●	
Spatial planning	●			●		

Icons made by Freepik, itim2101, Eucalyp, Icon Pond, Roundicons, from www.flaticon.com
COASTAL Deliverable 03: Sectoral Analysis of Coastal and Rural Development; Direct contributions from MAL leaders following interim consultation

emissions of nutrients still exceed the EU WFD target levels and contribute to coastal eutrophication. In addition, the quantity of fresh water is under pressure during extended periods of drought, as a result of multiple demands from industry, tourism, population and agriculture. A major stressor is the increasing salinization of inland waters, as a result of human waterworks, water management and sea-level rise. A main challenge for this case study is the fragmentation of policy and knowledge for coastal and rural development. A common administrative framework for coastal-rural integration is lacking, and policy responsibilities are fragmented at the regional and national level.

7.2.2.2 South-West Messinia (Eastern Mediterranean Region)

Agriculture (mainly olive farming) and coastal tourism are the two major economic activities in Western Messinia, Greece. Tourism is expanding and goes hand in hand with infrastructure development (hotels, roads and airports) and can provide opportunities for diversified livelihoods, but also increases pressures on the environment and cultural sites. Coastal areas are also affected by agrochemicals, soil erosion, solid waste landfills and waste water from inland sources. In particular, waste products from olive production constitute a threat to surface and coastal water quality. Climate change is expected to increase coastal erosion and decrease the availability of fresh water, with increased risk for saltwater intrusion into coastal wetlands and aquifers. There are also plans for offshore oil and gas exploration that will have implications for the area's rich coastal biodiversity. The study area comprises several important cultural sites and Mediterranean habitats included in the reference list of the Natura 2000 initiative.[9] The MAL will develop a number of alternative strategies for local economic development that will allow a diversification and strengthening of a sustainable local economy while minimizing the impact on the Natura 2000 sites. Long-term planning for sustainable tourism and agriculture will take into account resilience to future climatic changes, exploiting the expertise and experience of local stakeholders.

7.2.2.3 Norrström River Basin: Baltic (Baltic Sea Region)

The Norrström drainage basin (22,000 km^2) comprises the major part of the Swedish water management district Northern Baltic Proper. For the case study, economic and environmental issues are considered both at the local scale and that of the Baltic Sea, including environmental and physical interactions between the two scales. The basin is part of the fertile Swedish belt, characterized by extensive agriculture, and includes the Swedish capital Stockholm. An unresolved and well-recognized problem for the coastal environment and its sustainable development is the human-driven eutrophication and associated hypoxia with recurring harmful algae blooms, caused by the combined nutrient emissions from households, agriculture and industry. These impacts occur on the local/regional scale of the Swedish Norrström case and the macro-regional/cross-boundary scale of the whole-Baltic case. Sustainable development solutions must consider the agricultural, urban and industrial activities in the hinterland, the high and increasing population density, tourism development in the coastal zone, ongoing and future climate change and their combined effects on the nutrient loads transported to the coast. The overarching problem and possibilities for addressing the problem are recognized in strategic planning and development plans at the local and regional scale.

[9]https://ec.europa.eu/environment/nature/natura2000/index_en.htm

7.2.2.4 Charente River Basin (Atlantic Coast)

The Charente river basin in SW France is predominantly rural and covers an area of 10,500 km^2 linked to the sea by the Pertuis Charente River with a large tidal influence. Agriculture represents the main hinterland activity, and major urbanization and industrialization can be found at Port Atlantique-La Rochelle (8.4 million ton per year). The region is characterized by a contrast between the densely populated coastal fringe and the rural territory with a low population density. Industrial activity is dominated by agri-food and wine industry around the cities of Cognac and Angoulême. Tourism in the region is significant. At the river mouth, the Charente River supplies fresh water to the oyster basin of Marennes-Oléron. Fresh water is essential for shellfish farming, which depends on a sensitive balance between the temperature, salinity and acidity of coastal waters and inland supply of fresh water and nutrients. The Marennes-Oléron bay is the first European shellfish farming centre for spat production and for the number of shellfish companies (SMEs). Charente-Maritime is home to three major marshes, the cultivation of which poses a risk of imbalance, thus endangering the fauna and the flora. There is a high concern for diffuse pollution by nutrients and pesticides of surface water and their influence on the drinking water supply. Most inland watercourses suffer from droughts and periodic low water levels, making water supply a challenge. Thus, the main policy issue for the region is the current and future supply and quality of the fresh water for different functions (drinking water, agriculture, industry and shellfish farming).

7.2.2.5 Danube Mouth: Black Sea (Black Sea Region)

Due to the semi-enclosed location and size of the contributing catchment area, the Black Sea is vulnerable to anthropogenic pressures and pollution sources. The nutrient regime of the Danube has undergone significant changes due to increased economic activity, use of fertilizers, wastewater discharges and use of detergents, leading to changes in the Black Sea ecosystem. Eutrophication results in decreased transparency, higher quantities of organic matter decomposition and oxygen depletion with bottom waters becoming seasonally hypoxic or even anoxic. Since the early 1990s, decreasing nutrient inputs has resulted in signs of recovery. Today, the Black Sea catchment is still under pressure from excess nutrients and contaminants due to emissions from agriculture, tourism, industry and urbanization in the Danube basin. This hampers the achievement of a Good Environmental Status by 2020, as required by the EU-MSFD. The increased rates of eutrophication, pollution and bioaccumulation affect both the biodiversity (including Natura 2000 sites) and fishing sectors. Mass tourism is an important growth sector for the Black Sea and ecotourism is becoming more important in the region. Approximately 65% of the Romanian coastline is located in the Danube Delta Biosphere Reserve and subject to tourism regulations, resulting in conflicts between nature conservation and economic development. Failing to resolve these conflicts has economic and political impacts.

7.2.2.6 Mar Menor Coastal Lagoon (Western Mediterranean Sea)

The Mar Menor coastal lagoon (135 km^2) is located in the Region of Murcia (SE Spain). The area is characterized by multiple environmental, sociocultural and economic interests, often competing for scarce resources, water being the most important. There is a high potential for complementarity, win-win scenarios and development of sustainable business cases based on public-private collaboration, efficient use of water and innovative farming practices and a transition to sustainable models of tourism and agriculture. The catchment draining into the Mar Menor covers an area of 1.255 km2 and is mainly covered by intensive irrigated agriculture (35%) and tree crops (30%). The intensive and highly profitable irrigated agriculture depends on scarce low-quality groundwater and water from inland inter-basin water transfers. Agriculture provides labour and income to the region but forms a source of excessive nutrient and contamination into the Mar Menor coastal lagoon. The resulting poor water quality affects the ecology of the lagoon with severe implications for its potential function for tourism and fisheries. The coastal lagoon forms part of a Specially Protected Area of Mediterranean Importance (SPAMI). The Mar Menor is one of the hotspots for tourism in the Region of Murcia, with a total number of 346,000 tourists and 1.4 million overnight stays in 2016. Besides international visitors, the Mar Menor has an important touristic function for the regional population (1.5 million inhabitants). The availability of water for irrigation and drinking water for tourism will be further reduced under future climate conditions. As such, the Mar Menor is strongly influenced by interactions between inland agriculture on the one hand and coastal tourism and fisheries that affect natural ecological values and socioeconomic sustainability on the other hand. The need to move towards sustainable modes of agriculture, fishery and tourism is increasingly recognized and recently revived strongly due to sudden increase in contamination levels resulting in a considerable decline in tourism.

7.2.3 Embedding Stakeholder Perspectives

The general methodology of COASTAL is based on a complete integration of a participatory, multi-actor approach with evidence-based analysis, exploiting local and scientific knowledge and expertise. As such, local stakeholders from coastal and rural areas along with scientists work together to codesign innovative business solutions and policy alternatives to create sustainable, dynamic and resilient European land-sea regions. Stakeholders, entrepreneurs and administration representatives actively and jointly contribute to the design, testing and interpretation of these business and policy solutions.

A Collaborative System Dynamics[10] approach is used to provide a natural framework for describing and analysing the time-dependent development of complex systems such as the land-sea ecosystem, at a strategic level appropriate for business and policy support. The Multi-Actor Labs engage local stakeholders in codesigning metal maps of their regional land-sea interactions during interactive sectoral and cross-sectoral workshops.

Moreover, the MALs have five specific objectives:

- Facilitate and coordinate the exchanges between local coastal and rural stakeholders, policy makers, economic actors and stakeholders, aimed at conceptualizing the land-sea interactions and system feedback.
- Exploit the available scientific knowledge and existing data to quantify the system interactions, using System Dynamics (SD) modelling and Systems Thinking techniques.
- Contribute to the design, testing and demonstration of generic and reusable tools and key performance indicators for coastal and rural development and land-sea synergy.
- Develop practical, feasible and evidence-based business road maps and policy guidelines aimed at improving the land-sea synergy and collaborations between coastal and rural operators.
- Organize the local dissemination actions and inter-case study exchanges and contribute to the post-project exploitation of the project outcomes (Fig. 7.3).

Each local case study organizes six sectoral workshops from coastal and rural areas and one multisectoral workshop regrouping stakeholders from coastal and rural areas. At the final stage of the project, an international workshop will be held, bringing together local stakeholders from each case study. The main purpose of the workshops is to engage stakeholders in an open discussion, aimed at identifying the main issues, opportunities, obstacles and solutions in the context of land-sea interaction and their own sector or field of expertise (tourism, farming, water management, spatial planning). The mental mapping refers to the graphical representation of the issues brought forwards by the workshop participants, linking the elements mentioned. Causality between key variables is an important aspect of the mental maps (sometimes referred to as 'mind maps') (Tiller 2019). Additionally, local stakeholders give feedback on potential business opportunities and best practices applied with success in other coastal-rural regions and are engaged in the co-construction of scenarios and transition pathways to reach sustainability in the context of their respective case study areas (Fig. 7.4).

[10]System Dynamics modelling is a well-established methodology for analysing the behaviour of complex environmental and social systems, and understanding the counter-intuitive responses to business and policy decisions, resulting from the underlying balancing and reinforcing feedback structures.

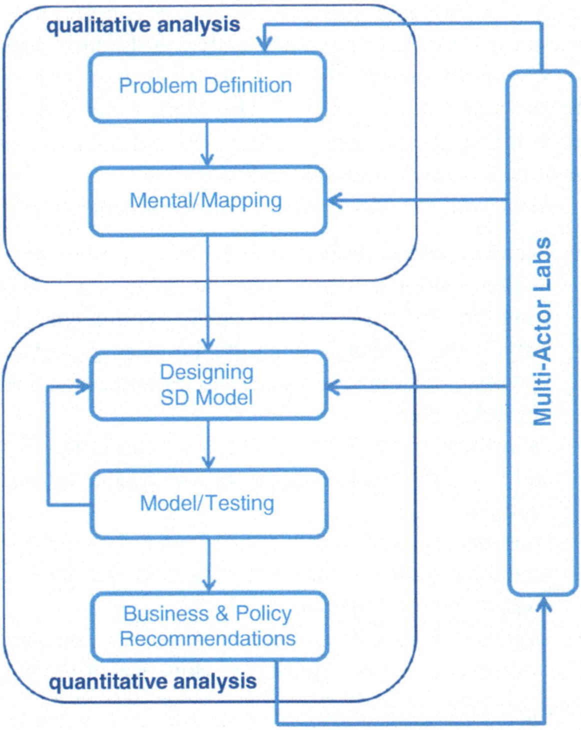

Fig. 7.3 Collaborative system dynamics modelling in COASTAL (Cf. COASTAL project)

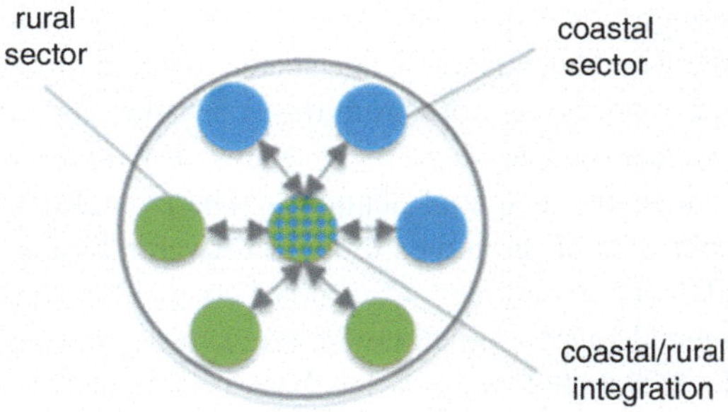

Fig. 7.4 COASTAL multi-actor labs (cf. COASTAL project)

7.2.4 *Fostering Land-Sea Synergies: Initial Findings from the COASTAL Project*

Following the sectoral workshops, the main issues facing the different coastal-rural stakeholders were identified. Despite the diversity of the case studies, it was possible to distil several common issues that need to be addressed.

As presented in Table 7.1, issues related to the water resource (regarding the quality and the scarcity of the resource), land availability as well as the lack of cooperation and conflicts between stakeholders are common to all case studies. In most of the COASTAL case study areas, the natural environment and its biodiversity are under threat, and local authorities face difficulties related to the protection and the management of these natural areas. The degradation of soil quality often related to salinization is another concern. Finally, these regions have to deal with issues related to the seasonality of tourism activity, waste management and coastal risk management (sea-level rise, flood risk, coastal erosion). Climate change is identified in all COASTAL case studies as a serious concern with the potential to exacerbate these issues and create additional impacts on all sectors.

This begs the question: how can developing land-sea synergies provide solutions to rural-coastal-sea issues? As demonstrated by the philosophy of the source-to-sea concept, stakeholders from land and sea regions share the same natural resources; mismanagement of inland natural ecosystems will inevitably affect ecosystem and human activities related to these resources on the coast and at sea. Actors within agriculture, energy, fisheries, cities, infrastructure and water management must all be collectively engaged as stakeholders in planning, operations and management from source to sea to protecting the natural resources and their related ecosystem services and ensuring a sustainable development of the land-sea regions.

As part of the COASTAL project, an inventory of best practices, successes and lessons learned related to business opportunities and policy alternatives has been compiled. The inventory provides examples from all over Europe and beyond, serves as a reference point for inspiring solutions to increase land-sea synergies and coastal-rural collaborations (Akinsete et al. 2019). However, practices with the main purpose of improving coastal-rural collaborations and land-sea synergies are still few and far between and are not yet commonplace. In fact, the sector workshops of the COASTAL project are one of the first attempts to identify the opportunities for land-sea synergy and added value for specifically for rural development. Nevertheless, a few other EU-funded projects addressed this particular challenge in a systematic way (for instance, with the C-Scope project which aims at achieving an integrated approach to management and planning within the land-sea interface[11]) or for a different domain (the ROBUST project fosters the development of rural-urban synergies[12]).

The scientific literature when analysing the socio-environmental and economic land-sea interactions generally take a conceptual and academic approach with little practical examples on how to increase synergies and collaborations. Despite international efforts to promote the idea of a coordinated global approach, local recommendations, practices and projects still generally take a sectoral focus, which at most extends to consider either the rural, coastal or marine issues depending on the sector. Recently, more initiatives have attempted to develop coastal-marine synergies

[11]http://www.cscope.eu/en/

[12]http://rural-urban.eu/

(i.e. development of Pesca tourism,[13] aquaculture on land[14]), which could provide a potentially inspiring operational framework for the development of coastal-rural synergies. Nevertheless, many initiatives, by working on sustainable tourism, natural conservation and water issues, risk management and local redevelopment, unwittingly create some form of land-sea synergies and coastal-rural collaborations (see further details in Akinsete et al., COASTAL deliverable D09). Following the identification and analysis of the examples of best practice, key business opportunities, lessons learned and policy recommendations are highlighted below.

Combined activities are one of the best examples on how to create synergies between land and sea: Fishery and tourism are so far the best example of successful combined activities in Europe,[15] replicable across the continent with the proper legislative adaptation. Inviting tourists on fishing boats at sea, while developing fish markets and valorizing fishing culture and heritage on land, fosters sustainable fishing practices and at the same time raises awareness of consumers for a responsible consumption of sea products. These land-sea synergies aid in the protection of the resource at sea while sustaining the entire fishing community on land. Other combined activities can be envisaged; however, despite extensive research and promising results (e.g. EU MERMAID, H2OCEAN and TROPOS projects for multi-use of offshore platforms, COEXIST[16] and MUSE[17] projects for combined activities at sea (i.e. between fishery and aquaculture; energy production and aquaculture or fishery; aquaculture and environmental protection; tourism, fishery and environmental protection)), the possibility of other combined activities at sea is still in the prototype stage in most cases and deeply dependant on the local context and legislations.

Ecotourism and agro-tourism based on local natural and cultural heritage can also create coastal-rural collaboration and land-sea synergies. Economic activities, population and tourists are concentrated in the coastal area, while the rural hinterland is often neglected but preserved from landscape degradation which in turns makes it attractive for alternative forms of tourism. In collaboration with coastal stakeholders, ecotourism and agro-tourism can create new economic opportunities in rural hinterland. It can also be used as a tool for reducing touristic pressure on the seaside area and extending the touristic season of a coastal-rural region, as such beneficial for both areas and creating land-sea synergies. In the Algarve region (Portugal), local initiatives promote an alternative tourism based on local products and heritage

[13]Combined activity of tourism and fishery, initially developed in Italy and now spreading in other Mediterranean countries (France, Spain) – cf. TourismMed interreg project https://tourismed.interreg-med.eu/

[14]Cf. Coastal Laboratory in the Netherlands – https://www.kustlaboratorium.nl/aquacultuur

[15]See examples of successful implementation of tourism-fishery combined activities in https://webgate.ec.europa.eu/fpfis/cms/farnet/files/documents/FARNET_Fisheries_and_Tourism-9_EN.pdf.

[16]http://www.coexistproject.eu/coexist-results/coexist-case-studies

[17]https://muses-project.com/

(establishment of a route of Traditional Salt from the Atlantic, the 'KM 0' branding initiative, the Eating Algarve Food tour cf. Akinsete et al. 2019).

The development of renewable energies offshore, if integrated as part of wider strategic development inland, can foster land-sea synergies as shown in the Renewable Energy Island project[18] of Samsø in Denmark. Through the development of renewable offshore and inland energies and the direct investment of the local community, the island became 100% self-sufficient in terms of energy production. Local islanders who are shareholders in the different energy production enterprises even generated income. The multiplier effect of the renewable energy strategy has, therefore, been felt throughout the local economy. Moreover, the project had positive effects of the quality of air, water and terrestrial environment with significant reductions in emissions of greenhouse gases and airborne particles, which has in turn benefited the provision of ecosystem services. Such an initiative is a perfect example on how the green economy and the needed energy transition towards renewables can create business opportunities, foster collaborations in a coastal-rural community and improve land-sea synergies which will benefit the economy, the environment and the local community. The development of offshore energy can also boost local port/shipping activities by creating new strands of income-generating activity and thus diversify the local economy as foreseen in the French COASTAL case study for the port of La Rochelle.

Coastal risk management strategies can also create opportunities for renewable energy development; flood risk and coastal defence systems can be combined with energy production systems at sea with break water-wave energy converters (Emid 2014). In the context of climate change, sea-level rise, increase deterioration of weather conditions indicating stronger and more frequent storms along the coasts, increases the risk of major disasters in coastal areas. Technological innovations in the form of breakwaters installed offshore have the ability to reduce the strength of sea waves and keep the sea surge off the coast, thus limiting flood risk and coastal erosion. These breakwaters can be coupled with wave energy converters to take advantage of the wave power and harness the energy for productive purposes. Although this technology is still at the testing phase, it is very promising in terms of coastal risk management and renewable energy development. It is also a good example of how Green Growth and Blue Growth may be coupled to foster land-sea synergies.

For coastal defence, nature-based solutions are also particularly relevant: beach nourishment,[19] compared to conventional options, creates additional co-benefits through increased attractiveness for recreational opportunities. Increasing the width of the beach may provide opportunities for the development of new recreational activities on the beach. It also generally increases the environmental quality and consequently the number of tourists. Such co-benefits can increase economic

[18] Cf https://energiakademiet.dk/en/

[19] Practice in which sediment is brought onto a beach to replace sediment lost through erosion process

activity, generating tax revenue, which in turn leads to leveraging of the overall public investments in the project. As a result of positive environmental and economic positive side effect outcomes, the beach nourishment solution has become economically more attractive than conventional solutions (De Bel et al. 2018). It is a good example on how a flood risk and coastal defence management project through nature-based solutions became an opportunity for cross-sectoral collaboration and business development.

Besides offering sustainable solutions to anthropogenic pressures, nature-based solutions for environmental issues in coastal and rural areas can also provide additional benefits in terms of landscape restoration and creating new avenues for recreational activity. In coastal-rural areas, it can create the basis for a circular economy through water recycling and reuse – as an alternative to expensive conventional water treatments, biological treatment using plants to absorb nutrients is, for instance, highly effective in the tropics where plants grow year round; however this option is often ignored by sewage designers from temperate zones, where plants only grow part of the year. Scientific findings also point out the advantages of land disposal for secondarily treated sewage effluence and wastewater reuse options. As an alternative to coastal discharges, areas with extensive wetlands could possibly be part of a secondary treatment/overland flow system, with the already nutrient-rich wetlands 'treating' the final effluent material. Also, instead of being expensively treated and ejected into the system, waste water could be reused, particularly by the agricultural sector which requires nutrient inputs or by coastal golf courses which are often in need of nutrient-rich effluent waters for irrigation to provide water and fertilizers at lower costs (see details in COASTAL Deliverable D09). A well-designed and planned water recycling and reuse system with nature-based solutions over the whole coastal-inland system (that includes stakeholders from coastal and rural areas in the design process) would reduce the impact of inland polluted water onto the coastal-sea ecosystems and by extension the impact on activities such as fishery, aquaculture and coastal tourism, which are dependent on good seawater quality.

Finally, a tool such as a spatial planning management strategy designs above the rural-coastal-marine boundaries can foster land-sea synergies and improve coastal-rural collaboration by developing a common vision for the whole area from the rural hinterland to the sea. This will allow the creation of a coherent and sustainable development plan for the future by taking into account rural-coastal-marine socio-environmental and economic interdependency. Such a strategy should be based on land-sea synergies and stakeholders' collaboration. However, the success of such a strategy will depend on the development of a global governance at the EU level, implying a deeper integration of the, still, siloed policies, i.e. between the WFD, the MSPD, the CAP and the ICZM approach which lack interconnections, as mentioned previously.

7.3 Conclusion

Increasing land-sea synergies is necessary in order to protect land and sea ecosystems, create sustainable and resilient coastal and sea regions and achieve the sustainable development goals. The COASTAL project seeks to increase land-sea synergies via an innovative dual approach using System Dynamics coupled with participatory methods, by involving local stakeholders from representative EU coastal regions to codesign business road map and policy alternatives. Initial work identifying main issues and business opportunities from representative case studies and the development of an inventory of best practice, successes and lessons learned regarding land-sea synergies and coastal-rural collaborations has yielded preliminary findings which indicate that new, innovative practices such as combined activities, alternative forms of tourism, the development of renewable energies offshore or coastal risk management strategies can play a role in creating a common vision for the development of an interconnected rural-coastal-sea region. Policies are still fragmented, but international initiatives (political and scientific) work towards unified, multisectoral framework and a better integration of the land and sea ecosystems. The source-to-sea concept highlights the gaps and needs for a better integration, cooperation and coordination of activities from the rural area to the ocean in order to achieve a harmonized, sustainable development of a land-sea area.
Funding: This research has received funding from the European Union's Horizon 2020 research and innovation program under grant agreement N° 773782.

References

Akinsete, E., Apostolaki, S., Guittard, A., Elwattar, S., Koundouri, P., & Tsani, S. (2019). *COASTAL deliverable D09: Inventory of business opportunities and policy alternatives*, COASTAL EU H2020 project, 117 p.

Bigagli, E., Pérez Valverde, C., Abdul Malak, D., & Guitart, C. (2019). *Tackling marine litter in the mediterranean: Knowledge and tools*, Policy report by the MED biodiversity protection community's working group "Biodiversity management and protection, Interreg Mediterranean, 40 p.

Borja, A., et al. (2010). Marine management – Towards an integrated implementation of the European marine strategy framework and the water framework directives article. *Marine Pollution Bulletin, 60*(12), 2175–2186.

Brown, S., Nicholls, R., Woodroffe, C., Hanson, S., Hinkel, J., Kebede, A. S., et al. (2013). *Sea-level rise impacts and responses: A global perspective* (pp. 117–149). Dordrecht: Springer.

Degeorges, A., Goreau, T., & Reilly, B. (2010). Land-sourced Pollution with an emphasis on domestic sewage: Lessons from the Caribbean and implications for coastal development on Indian Ocean and Pacific coral reefs. *Sustainability., 2*, 2919–2949. https://doi.org/10.3390/su2092919.

De Bel, M., Kok, S., & Hinkel, J. (2018). D5.1 policy brief: Leveraging public finance for coastal adaptation, *Green-Win project*, H2020, 10 p. Retrieved from https://green-win-cloud.org/index.php/s/jHH9sDV1TqHUzT3#pdfviewer

Emid, S. (2014). *BREAKWATER – WAVE ENERGY CONVERTER Coastal defence and cheap evergreen energy production, input paper*. Prepared for the Global Assessment Report on

Disaster Risk Reduction 2015, UNISDR, GAR 9 p. Available: https://www.preventionweb.net/english/hyogo/gar/2015/en/bgdocs/inputs/Emid,%202014.%20breakwater%20wave%20energy%20converter.pdf. Accessed 13 June 2019.

Erban, L. E., Gorelick, S. M., Howard, A., & Zebker, H. A. (2014). Groundwater extraction, land subsidence, and sea-level rise in the Mekong Delta, Vietnam. *Environmental Research Letters, 9*. 084010. (6pp). https://doi.org/10.1088/1748-9326/9/8/084010.

Granit, J., Liss Lymer, B., Olsen, S., Tengberg, A., Nõmmann, S., & Clausen, T. J. (2017). A conceptual framework for governing and managing key flows in a source-to-sea continuum. *Water Policy, 00*, 1–19.

Halpern, B. S., Walbridge, S., Selkoe, K. A., Kappel, C. V., Micheli, F., D'Agrosa, C., Bruno, J. F., Casey, K. S., Ebert, C., Fox, H. E., Fujita, R., Heinemann, D., Lenihan, H. S., Madin, E. M. P., Perry, M. T., Selig, E. R., Spalding, M., Steneck, R., & Watson, R. (2008). A global map of human impact on marine ecosystems. *Science, 319*, 948–952. https://doi.org/10.1126/science.1149345.

Howarth, R. W., Sharpley, A., & Walker, D. (2002). Sources of nutrient pollution to coastal waters in the United States: Implications for achieving coastal water quality goals. *Estuaries, 25*, 656–676. https://doi.org/10.1007/BF02804898.

Le Moal, M., Gascuel Odoux, C., Ménesguen, A., Souchon, Y., Étrillard, C., Levain, A., Moatar, F., Pannard, A., Souchu, P., Lefebvre, A., & Pinay, G. (2019). Eutrophication: A new wine in an old bottle? *Science of the Total Environment, 651*(Part 1), 1–11.

Li, Y., Westlund, H., & Liu, Y. (2019). Why some rural areas decline while some others not: An overview of rural evolution in the world. *Journal of Rural Studies*. https://doi.org/10.1016/j.jrurstud.2019.03.003.

McGranahan, G., Balk, D., & Anderson, B. (2007). The rising tide: Assessing the risks of climate change and human settlements in low elevation coastal zones. *Environment and Urbanization, 19*, 17–37.

Nilsson, C., Reidy, C. A., Mats Dynesius, M., & Revenga, C. (2005). Fragmentation and Flow Regulation of the World's Large River Systems, Report. *Science*. 15 Apr 2005, *308*(5720), 405–408. https://doi.org/10.1126/science.1107887.

Small, C., & Nicholls, R. J. (2003). A global analysis of human settlement in coastal zones. *Journal of Coastal Research, 19*, 584–599.

Smith, K. (2011). We are seven billion. *Nature Climate Change, 1*, 331–335.

Syvitski, J. P. M., Kettner, A. J., Overeem, I., Hutton, E. W. H., Hannon, M. T., Brakenridge, G. R., et al. (2009). Sinking deltas due to human activities. *Nature Geoscience, 2*, 681–686.

Tiller, R. (2019). *COASTAL: Sectoral analysis of coastal and rural development*, Deliverable D03, EU H2020 COASTAL project, 124 p.

UNEP/GPA. (2012). *Manila declaration on furthering the implementation of the global programme of action for the protection of the marine environment from land-based activities*. United Nations Environment Programme (UNEP). Global Programme of Action for the Protection of the Marine Environment from Land-based Activities (GPA). Available at: http://www.unep.org/regionalseas/globalmeetings/15/ManillaDeclarationREV.pdf

Chapter 8
Monetary and Non-monetary Valuation of Cultural Ecosystem Services in Marine Protected Areas

Lydia Stergiopoulou, Phoebe Koundouri, and Achilleas Vassilopoulos

Abstract Cultural Ecosystem Services (CES) are recognized but still considered as the "residual" ES subcategory and remain understudied. Their potential to shape common identities and impact societal perspectives on ocean/marine resources' management explains why further research on CES can widen the range of information needed for policymaking, especially in cases of blue tourism interventions. In this chapter (This work has received funding from the European Union's Interreg Balkan-Mediterranean programme under grant agreement MIS 5017160.), we review some possible conceptual frameworks for the CES classification along with the monetary and non-monetary (revealed and stated preference) methods for their valuation. Attention is given to the stated methods that the last years have received increasing attention and exhibit some potential to be linked with Maritime Spatial Planning decisions. An attempt to operationally define CES in the context of Marine Protected Areas and investigate the determinants of perceived cultural heritage and identity features has been adopted in two Interreg projects, AMAre and RECONN ECT.

Keywords Marine protected areas · Cultural ecosystem services · Non-monetary methods · Maritime spatial planning

---

L. Stergiopoulou (✉)
United Nations Sustainable Development Solutions Network, Athens, Greece
e-mail: lydia.stergiopoulou@icre8.eu

P. Koundouri
School of Economics and ReSEES Laboratory, Athens University of Economics and Business, Athens, Greece

United Nations Sustainable Development Solutions Network Europe, Paris, France

EIT Climate–KIC Hub Greece, ATHENA Research and Innovation Center, Athens, Greece

ICRE8: International Centre for Research on the Environment and the Economy, Athens, Greece

A. Vassilopoulos
Department of Economics, University of Ioannina, Ioannina, Greece

P. Koundouri (ed.), *The Ocean of Tomorrow*, Environment & Policy 57,
https://doi.org/10.1007/978-3-030-56847-4_8

8.1 Introduction

The concept of Ecosystem Services (ES) originates in the 1970s when a first attempt to define the beneficial ecosystem processes and functions was made based on a utilitarian approach, aiming to increase public support for ecosystems' conservation. In the mid-1990s, the economists Costanza and Daily (Daily 1997) were among the first researchers to introduce in the literature the concept of ecosystem services valuation (Costanza et al. 1997). During the 2000s, the concept of ES has already received much attention, and after the publication of the Millennium Ecosystem Assessment synthesis reports (MEA 2005), it is well-placed in the policy agenda. In MEA, the Cultural Ecosystem Services (CES) are for the first time officially defined as a framework for understanding the ecosystems' processes and functions, as well as the relationship of ecosystems with human well-being. Long before that, attempts to integrate the economic aspects of ES within policy decision-making have been made and resulted in the establishment of several markets of ecosystem services or payment-for-ecosystem-services schemes within that context. However, the CES is the least developed category of ES, especially as far as the context of "ocean space" is concerned.

Later, another framework that has also attracted the interest of the research community, The Economics of Ecosystems and Biodiversity (TEEB 2010), explicitly included CES as a services category linked to commercial or non-commercial cultural-related values. But more than a decade later, CES are still not fully operationalized in valuation exercises and the decision-making process, while the definitions, conceptual models and assessment indicators related to CES are not yet standardized. The qualitative and interpretative nature of CES as well as their confusing and overlapping meanings can explain why standard assessment indicators are missing and why their measurement raises methodological challenges.

The boundaries of Marine Protected Areas (MPAs) contain ocean/marine space or settings that can combine several cultural elements such as protected shipwrecks, marine conservation areas, marine parks, settlements, green spaces for leisure, diving sites, etc. Thus, MPAs have the potential to bring new levels of integration and protection to the management of cultural resources within the coastal and marine environments. This explains how cultural benefits are placed within the context of MPAs, which can be associated with a range of culturally defined attributes (e.g. scenic beauty, distinctiveness, etc.). Although management plans based on the valuation of ecosystem services provided by the habitats within their boundaries have a track record of application (Börger et al. 2014), CES are rarely part of the valuation exercises because they take place at different levels and realized by diverse methods, as will be explained in the next sections.

8.2 Operationalization of CES

A literature review by Cheng et al. (2019) on the CES valuation methods in 293 publications showed that the rate of publications increased from 3 papers per year in 2005 to 90 papers per year until 2017. This fact reveals the increasing need to operationally define CES in socio-economic models. The multitude of perspectives for approaching CES reflects a new multidisciplinary dynamic field under progress, but it also shows a lack of well-established and readily applicable research framework. Different conceptual frameworks have been used for the categorization and definition of the CES, attempting to find an approach that can render the CES operational, not only in the research but also in policy agenda and in decision-making.

The Millennium Ecosystem Assessment first defined ES as the "benefits that people obtain from ecosystems" (MEA 2005) and divided them in provisioning, regulating and supporting cultural services. In MEA, CES are defined as mostly intangible and invisible that related to the "non-material benefits that people obtain from the ecosystem" and influence the quality of life, while the provisioning, regulating and supporting ES are considered as material services. In the same framework, the classification is based on the practices and experiences that arise in the environmental spaces under consideration, and for that reason, it has been criticized in various occasions. The notion behind the critique is that some practices such as recreation (consumptive or non-consumptive) fall into two services categories (e.g. cultural and provisioning), a fact that may result in double-counting.

Another classification is the Common International Classification of Ecosystem Services (CICES) developed in 2013 by the European Environment Agency (EEA) as part of the work on environmental-economic accounting, which is led by the United Nations Statistical Division. Its central focus is the standardization of the ES description to ease their accounting. Besides ES valuation, it introduces the need of mapping. CICES divides all ES in classes, where each ES is made up of a biophysical output and an eventual use or benefit. In CICES, services are distinguished from benefits and CES are defined as "the characteristics of elements of nature that provide opportunities for people to derive cultural goods or benefits". Further CES subcategorization depends on whether the opportunities are realized from direct contact or remote interaction with nature (Haines-Young and Potschin 2018). This classification became the link with the Mapping and Assessment of Ecosystem and their Services (MAES), a flagship project of the EU which proposes specific indicators for the measurement of the condition of terrestrial, freshwater and marine ecosystems (European Commission 2018).

A third classification system is the UK National Ecosystem Assessment and its follow-on phase (UK NEAFO) that categorizes values according to the environmental settings or spaces were CES can be identified. Specifically, the system mentions that CES "encompass the environmental spaces and cultural practices that give rise to a range of material and non-material benefits to human well-being (Jobstvogt et al. 2014). These spaces and practices interact with contemporary cultural values to shape people's identities, provide experiences that contribute benefits in terms of well-being, mental and physical health, and equip people with a range of skills and capabilities" (Fish et al. 2016). This framework allows for an assessment of CES taking into consideration the extent and state of different spaces, the practices (e.g. snorkeling, wildlife watching, diving) and the well-being benefits associated with spaces and practices through capacities (e.g. knowledge and skills), identities (e.g. cultural identities) and experiences (e.g. spiritual and aesthetic experiences). For the above reasons, that framework can allow the consideration of multiple cultural and natural features in the geographical area under consideration and thus is also suitable in MPAs.

8.3 Monetary and Non-monetary Valuation Methods

An approach used for describing and capturing the benefits from the diverse ES is the Total Economic Value (TEV) (Defra 2007), which is based on individuals' preferences extracted either via stated or via revealed preference methods. To link the TEV with the designation and management of MPAs with regard to the CES, we need to identify the values categories within the various ecosystems of an MPA. Under this framework, examples of direct use CES values may be those attached to species watching, snorkeling, diving, boating, angling, etc. within the MPA ecosystems. Regarding indirect use CES values, examples might include values derived through research, education, etc. Non-use bequest and existence CES values may come from MPA wilderness and seascape preservation for the next generations and the appreciation of the place identity created by the MPA ecosystem, even if never visited. Figure 8.1 depicts the above classification (Remoundou et al. 2014).

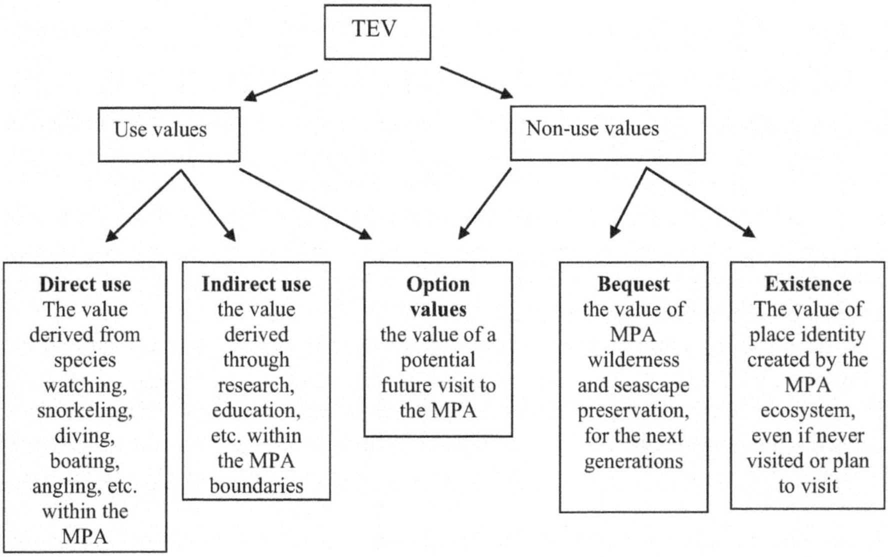

Fig. 8.1 Total economic value of CES-related MPA ecosystems

The methods to estimate the economic values described above are divided into two categories, revealed and stated preference approaches; both can be monetary and non-monetary. In the case of monetary methods, the revealed preference methods are based on market prices or observed choices (e.g. touristic destinations) as proxies of public's preferences over ecosystem services.

The stated preference methods, on the other hand, are based on the construction of hypothetical markets and are conducted through direct questionnaires. In such questionnaires, the public/stakeholders are asked to directly place a value to the ES described in the survey or to choose between hypothetical scenarios that include bundles of services affected by the management scenarios. The services are described by unique characteristics of each ecosystem (or "attributes") which in the case of CES can be leisure features of the area or other features for which tourists and locals are attaching a value on protecting against future degradation/harm. These characteristics are combined with a cost attribute (usually MPA entrance fee of tax) to form scenarios or choice situations, and then researchers can infer their relative rankings and monetary values by collecting preference data.

During the last few years, the strand of literature that involves non-monetary valuation methods is receiving increasing attention. Cheng et al. (2019) identify 13 such non-monetary methods, that they also divide in revealed and stated preference, just like the monetary ones discussed above. The three revealed methods use observation, and/or images, photos posted by the public, advertisements or data from social media (that might include underwater cultural heritage, marine wildlife, extraordinary seabed, etc.) as proxies to obtain the values on people's preferences. Stated preference methods use again questionnaires, but unlike monetary valuation, the questions now focus on qualitative information. Such narratives are obtained by asking respondents to either engage in storytelling or describe their perception of well-being from CES, using various indicators. These indicators can be combined with other techniques, such as participatory mapping, participatory GIS or public participatory GIS to combine citizen science and inform spatial planning decisions. The "participatory mapping method" links the CES perceived by respondents to a specific location, while the participatory GIS incorporates GIS, GPS and remote sensing image analysis software with interviews/questionnaires and human spatial knowledge. Finally, the public participatory GIS brings out the perceptions of the public with the use of geographic technology education.

8.4 CES Valuation and MPA Management

8.4.1 Monetary Valuation

The EU Maritime Spatial Planning Directive 2014/89/EU defines the Maritime Spatial Planning (MSP) as "a process by which the relevant Members State's authorities analyse and organize human activities in marine areas in order to achieve ecological, economic and social objectives". In its turn, it follows the ecosystem-based approach, meaning that MSP should be based on the best available scientific knowledge on the ecosystems and their dynamics. Its practical implementation necessitates, among others, the evaluation of conflicting uses and interests and states that the marine strategies "shall apply an ecosystem-based approach to the management of human activities, ensuring that the collective pressure of such activities is kept within levels compatible with the achievement of good environmental status and that the capacity of marine ecosystems to respond to human-induced changes is not compromised, while enabling the sustainable use of marine goods and services by present and future generations".

As part of the MSP process, MPAs should be able allocate marine resources to anthropogenic uses and integrate geospatial and scientific information in the decision-making. MPA designation and management should aim to identify and map areas that are useful for each use and then minimize the conflicts between ecological, social and economic interests. The CES at the depths of MPAs can accrue from underwater seascape, seabed aesthetic, biodiversity, iconic and non-iconic species and archaeological remains (in ocean these are historic shipwrecks and rarely late Pleistocene/early Holocene remains), and the resulting benefits for the public are associated with history, heritage, education and identity related to the sea or to aesthetics. Although currently there is no EU policy to govern CES, the ES management approach has already had an impact on the policies regarding natural resources management. This is traced back to 1992, when the UN Convention on Biological Diversity (CBD) made the ES management approach the primary framework for action to achieve a balance among the three objectives of the Convention: the integrated management of land, water and living resources. In 2008, the ES management approach appears as a cornerstone of the EU Marine Strategy Directive (2008/56/EC) along with its amendment in 2017, which links it to the human pressures and impacts on the marine environment, using the 11 descriptors and the aim of achieving Good Environmental Status of the EU's marine waters by 2020 (Marine Strategy Framework Directive 2008).

To make CES values operational in MPA management using a monetary stated preference approach, we need to define the attributes that can be used for their monetary valuation as described in the previous section. A pilot operationalization of the proposed methodology was developed in the case of the Interreg projects REC ONNECT (Regional Cooperation for the transnational ecosystem sustainable development) and AMAre (Actions for Marine Protected Areas). In these projects, the essential socio-economic and cultural variables offered by seagrass habitats of targeted MPAs in Greece, Cyprus, Albania and Bulgaria (RECONNECT) and Spain, Malta, Italy and Greece (AMAre) were studied, with the overall objective to develop management scenarios based on the seagrass ES valuations that can be of use for the policymakers. The CES included were relevant to the natural environment itself, in particular, the ability of *Posidonia* meadows to become hot spots for biodiversity providing food, habitat, refuge and nursery ground for marine flora and fauna, including iconic or non-iconic species, and their ability to reduce current velocities and water turbidity, contributing to water clarity and purification. Consequently, the experimental design included attributes such as "aesthetic benefits" and "water visibility/clarity". Such attributes are particularly important for divers as well as for snorkelers and bathers who can enjoy the seascape beauty from the surface in low depths.

RECONNECT also included attributes accruing from the underwater cultural heritage, whose presence is preserved by seagrass. *Posidonia oceanica* has a particular function to lock out oxygen that otherwise degrades the archaeological remains and to form a protective matt above these treasures. As a result, the "preservation of underwater cultural heritage" was included in the design, assuming that the higher the number of archaeological remains per extent area, the more possibilities to capture cultural tourism preferences. The presence of underwater cultural spots of importance to visit (such as widely known shipwrecks) increases the tourist attraction even to non-divers.

Other parameters taken into account but not monetized in both projects were the underwater seascape (the more "beautiful" the underwater scenery that snorkelers and divers will come across, the higher is the possibility to be willing to visit or to preserve it), the diversity of substrate type (the wider the substrate, the more opportunities for bather/snorkelers/divers to visit the area more than once) and the spots with extraordinary fish productivity (the higher the fish stock, the higher the education opportunities). Other attributes are made up of existing diving routes (qualified tour operators and diving centres can showcase the MPA and increase the perceived value of diving. Routes that are tested and divided to levels of difficulty are more attractive to divers), level of diving difficulty (the deeper the scenery to visit (shipwrecks, reefs, tunnels, etc.), the less divers can access) and seabed quality (the existence of rock formations, archways, tunnels and vertical walls offers a very particular experience to snorkelers and divers).

8.4.2 Non-monetary Valuation

For non-monetary valuation of CES within the MPA framework, qualitative information may come from previous research related to sociocultural value of ecosystems, such as Bryce et al. (2016) and Schmidt et al. (2016). Bryce et al. (2016) presented Likert-scale indicators from known constructs, such as the reflection and sense of wholeness (Dallimer et al. 2012; Fuller et al. 2007; Irvine et al. 2010), the sense of place identity and continuity with the past (Dallimer et al. 2012; Fuller et al. 2007; Tengberg et al. 2012), the transformative values and inspiration (Chan et al. 2012a, b) as well as some newly suggested well-being indicators inspired by other scales, like the Human Scale Development Matrix (Cruz et al. 2009; Max-Neef 1989), the Monitor of Engagement with the Natural Environment (Natural England 2012) and the UK National Ecosystem Assessment of Cultural Services (Church et al. 2011).

A number of identified factors and corresponding indicators have been identified, such as the engagement and interaction with nature, the place identity and therapeutic value, captured by indicators showing how people connect with nature (e.g. educational or spiritual). Other indicators could be the ones used from Schmidt et al. (2016) for parks' benefits using CICES (Haines-Young and Potschin 2013) definitions. These include recreation, sense of place and inspiration perceptions, with the CES-related ones being either experiencing nature (experiential use of plants, animals and land- or seascapes in different environmental settings), physical use of nature (physical use of land- or seascapes in different environmental settings), education, cultural heritage or aesthetics.

8.5 Conclusions

The management of human activities in marine areas is particularly complex, due to the usually fragmented political and administrative nature of such areas. The intensive use of maritime space calls for more integrated management practices, to avoid negative effects on marine ecosystems, user conflicts, and to create synergies between maritime activities and promote the blue economy. MPAs can be an efficient tool to achieve these goals, and ES-based management has a vital role in this process, considering that in most marine areas, human activities are not spatially managed and monitored, while human impacts on ecosystem services are not taken into account when management initiatives are considered. But even in cases where ES do have a role in MPA management, CES are usually the least considered, if not ignored. However, the designation of MPAs that include historic shipwrecks, endangered habitats, etc. can offer combined opportunities for leisure activities and blue tourism but also social resistance by impacting people's perceptions on the marine environment and affect the CES such as seascape, leisure and cultural identities.

As a result, CES valuation can become an extremely useful tool that can shed light to the benefits derived from the cultural aspects of MPAs, guiding policymakers and management authorities. Integrated and adaptive management will help MPA managers to identify and adopt policies and practices that involve both cultural and natural resources at the ecosystem and landscape levels. This necessitates perception studies on CES to reveal potential conflicts and trade-offs or synergies which in turn highlights the need for a common framework and further research in monetary and non-monetary methods for capturing all the necessary information. So far, the variety of conceptual frameworks around the CES categorization has undermined this opportunity. Recent developments and the shift of the research agenda can create a fertile ground for the integration of CES in MPA planning and management. Given the highly subjective nature of CES, stated preference methods have the lion's share in this debate, and there seems to be a consensus that the more CES will become important, the more these methods will need to be developed to accommodate the specificities associated with these services. Recent projects that apply the combination of monetary and non-monetary valuation methods in MPA management are AMARE (Interreg) and RECONNECT (Interreg).

References

AMARE (Interreg). (2020). *AMARe – Actions for marine protected areas*. Online Platform. https://amare.interreg-med.eu

Börger, T., Beaumont, N. J., Pendleton, L., Boyle, K. J., Cooper, P., Fletcher, S., et al. (2014). Incorporating ecosystem services in marine planning: the role of valuation. *Marine Policy, 46*, 161–170.

Bryce, R., Irvine, K. N., Church, A., Fish, R., Ranger, S., & Kenter, J. O. (2016). Subjective well-being indicators for large-scale assessment of cultural ecosystem services. *Ecosystem Services, 21*, 258–269.

Chan, K. M. A., Satterfield, T., & Goldstein, J. (2012a). Rethinking ecosystem services to better address and navigate cultural values. *Ecological Economics, 74*, 8–18.

Chan, K. M. A., Guerry, A. D., Balvanera, P., Klain, S., Satterfield, T., Bostrom, A., Chuenpagdee, R., Gould, R., Halpern, B. S., Hannahs, N., Levine, J., Norton, B., Ruckelshaus, M., Russell, R., Tam, J., & Woodside, U. (2012b). Where are cultural and social in ecosystem services? A framework for constructive engagement. *Bioscience, 62*, 744–756.

Cheng X., Van Damme S., Li L., Uyttenhove P. (2019). 'Evaluation of cultural ecosystem services: A review of methods', Ecosystem Services, Elsevier 37.

Church, A., Burgess, J., Ravenscroft, N., Bird, W., Blackstock, K., Brady, E., Crang, M., Fish, R., Gruffudd, P., Mourato, S., Pretty, J., Tolia-Kelly, D., Turner, K., & Winter, M. (2011). Cultural services. In *UK national ecosystem assessment: technical report*. Cambridge: UNEP-WCMC.

Convention on Biological Diversity. (1992). *United Nations environment programme*, 1760 UNTS 79; 31 ILM 818.

Costanza, R., et al. (1997). The value of the world's ecosystem services and natural capital. *Nature, 387*(15), 253–260.

Cruz, I., Stahel, A., & Max-Neef, M. (2009). Towards a systemic development approach: Building on the human-scale development paradigm. *Ecological Economics, 68*, 2021–2030.

Daily, G. C. (Ed.). (1997). *Nature's services: Societal dependence on natural ecosystems*. Washington, DC: Island Press. 392 pp.

Dallimer, M., Irvine, K. N., Skinner, A. M. J., Davies, Z. G., Rouquette, J. R., Maltby, L. L., Warren, P. H., Armsworth, P. R., & Gaston, K. J. (2012). Biodiversity and the feel-good factor: Understanding associations between self-reported human Well-being and species richness. *Bioscience, 62*, 47–55.

Defra, U. (2007). *An introductory guide to valuing ecosystem services. Department for Environment, Food and Rural Affairs (Defra)*. UK: UK.

European Commission. (2018). *Mapping and Assessment of Ecosystems and their Services: An analytical framework for mapping and assessment of ecosystem condition in EU*, Fifth technical report

Fish, R., Church, A., & Winter, M. (2016). Conceptualising cultural ecosystem services: A novel framework for research and critical engagement. *Ecosystem Services, 21*, 208–217.

Fuller, R. A., Irvine, K. N., Devine-Wright, P., Warren, P. H., & Gaston, K. J. (2007). Psychological benefits of greenspace increase with biodiversity. *Biology Letters, 3*, 390–394.

Haines-Young, R., & Potschin, M. (2013). *Consultation on CICES version 4.3, August–December 2012: Report to the European Environment Agency. Consultation report on the common international classification of ecosystem services under EEA framework contract no EEA/IEA/09/003*. Nottingham, UK: Centre for Environmental Management, University of Nottingham.

Haines-Young, R., & Potschin, M. B. (2018). *Common International Classification of Ecosystem Services (CICES) V5.1 and guidance on the application of the revised structure.*

Irvine, K., Fuller, R., Devine-Wright, P., Payne, S., Tratalos, J., Warren, P., Lomas, K., & Gaston, K. (2010). Ecological and psychological value of urban green space. In J. Jenks & C. Jones (Eds.), *Dimensions of the sustainable city*. Dordrecht: Springer.

Jobstvogt, N., Watson, V., & Kenter, J. O. (2014). Looking below the surface: The cultural ecosystem service values of UK marine protected areas (MPAs). *Ecosystem Services, 10*, 97–110.
Marine Strategy Framework Directive (MSFD). (2008). Directive 2008/56/EC of the European Parliament and of the Council of 17 June 2008 establishing a framework for community action in the field of marine environmental policy. *Official Journal of the European Union, 164*, 19–40.
Max-Neef, M. (1989). Human scale development: An option for the future. *Development Dialogue, 1*, 5–81.
Millennium Ecosystem Assessment (MEA). (2005). *Ecosystems and human well-being: Synthesis*. Washington, DC: Island Press.
Natural England. (2012). *Monitor of Engagement with the Natural Environment: The national survey on people and the natural environment*. Available from: https://www.gov.uk/government/collections/monitor-of-engagement-with-the-naturalenvironment-survey-purpose-and-results
RECONNECT (Interreg). (2020). *Regional cooperation for the transnational ecosystem sustainable development*. Online platform. https://reconnect.hcmr.gr
Remoundou, K., Koundouri, P., Kontogianni, A., Nunes, P. A. L. D., & Skourtos, M. (2014). Valuation of natural marine ecosystems: An economic perspective. *Environmental Science and Policy, 12*, 1040–1051.
Schmidt, K., Walz, A., Jones, I., & Metzger, M. J. (2016). The sociocultural value of upland regions in the vicinity of cities in comparison with urban green spaces. *Mountain Research and Development, 36*(4), 465–474.
Tengberg, A., Fredholm, S., Eliasson, I., Knez, I., Saltzman, K., & Wetterberg, O. (2012). Cultural ecosystem services provided by landscapes: Assessment of heritage values and identity. *Ecosystem Services, 2*, 14–26.
The Economics of Ecosystems and Biodiversity. (2010). *Mainstreaming the economics of nature: A synthesis of the approach, conclusions and recommendations of TEEB, 2010*.

Chapter 9
Tackling Single-Use-Plastic Products in the Easter Mediterranean Sea: The BL.EU. Climate and MEDfreeSUP Projects

Phoebe Koundouri, Lydia Papadaki, Alice Guittard, Elias Demian, and Ebun Akinsete

Abstract The Mediterranean Sea is a top tourism destination in the world hosting more than 320 million tourists a year, but it's also one of the most affected areas by marine litter worldwide, polluting its shores and pristine coastal waters. Marine litter is estimated to cause an annual economic loss of €61.7 million to the EU fishing fleet because of reduced catch and damage to vessels, while polluted beaches can discourage tourists with consequent job losses in the sector. In this chapter (This work has received funding from the European Union's European Institute of Innovation and Technology under grant agreement N° 190880 and N° 200805), two projects funded by EIT Climate-KIC (2020) are being presented. The BL.EU.

P. Koundouri (✉)
School of Economics and ReSEES Laboratory, Athens University of Economics and Business, Athens, Greece

United Nations Sustainable Development Solutions Network Europe, Paris, France

EIT Climate–KIC Hub Greece, ATHENA Research and Innovation Center, Athens, Greece

ICRE8: International Centre for Research on the Environment and the Economy, Athens, Greece
e-mail: pkoundouri@aueb.gr

L. Papadaki
United Nations Sustainable Development Solutions Network Greece, Athens, Greece

EIT Climate–KIC Hub Greece, ATHENA Research and Innovation Center, Athens, Greece

A. Guittard
ICRE8: International Centre for Research on the Environment and the Economy, Athens, Greece

E. Demian
Foundation for Economic and Industrial Research (IOBE), Athens, Greece

E. Akinsete
ICRE8: International Centre for Research on the Environment and the Economy, Athens, Greece

United Nations Sustainable Development Solutions Network, Athens, Greece

EIT Climate–KIC Hub, Athens, Greece

P. Koundouri (ed.), *The Ocean of Tomorrow*, Environment & Policy 57,
https://doi.org/10.1007/978-3-030-56847-4_9

Climate project addressed the challenge of plastic marine littering in southern European waters by building capacity in three Mediterranean countries: Greece, Portugal and Croatia. The project is identifying the plastic marine littering issue at the very beginning of its life cycle and on the prevention side that can lead to plastic waste reduction and in consequence reducing carbon emissions from both production and waste management stages. The MEDfreeSUP project aims to set replicable voluntary protocols for free single-use plastics food packaging adoption for cafes, restaurants, foods stores, hotel, beach facilities but also public events and public places in three Mediterranean countries: Greece, Italy and Croatia. The project, which is ongoing, provides support and guidance to local business to comply with the EU SUP Directive and to engage Mediterranean islands and cities in the transition towards a free single-use plastic environment. This chapter presents the key findings and challenges of these projects dealing the impact of single-use plastics in Greece, which is one of the projects' countries.

Keywords Plastic pollution · Marine litter · Single-use plastics · Climate-KIC

9.1 The Challenge of Marine Litter in the Mediterranean Sea

Plastics are synthetic or semi-synthetic compounds made from carbon-based materials with specific properties, which are widely used because of their durability. Due to this, plastics are used widely also in the packaging of food products. Management of plastic waste is crucial, as most plastics are not biodegradable and remain in the natural ecosystems for hundreds of years. Today, only one third of the 27 million tonnes of plastic waste generated each year in Europe is actually recycled (WWF 2018).

The Mediterranean Sea[1] is a top tourism destination in the world hosting nearly 314 million international tourists a year, with European Mediterranean countries attracting most of the tourists, but it is also one of the most affected areas by marine litter worldwide, polluting its shores and pristine coastal waters (UNWTO 2014; UNEPMAP 2020). According to a WWF Report (2018), the Mediterranean Sea is today one of the seas with the highest levels of plastic pollution in the world, accounting for 95% of the waste in the open sea, on the seabed and on beaches across the Mediterranean. This waste comes mainly from the land and marine-related activities like fisheries, tourism and maritime transport. Marine litter consists of a wide range of materials, which vary regionally.

[1]The countries surrounding the Mediterranean in clockwise order are Spain, France, Monaco, Italy, Slovenia, Croatia, Bosnia and Herzegovina, Montenegro, Albania, Greece, Turkey, Syria, Lebanon, Israel, Egypt, Libya, Tunisia, Algeria and Morocco; Malta and Cyprus are island countries in the sea.

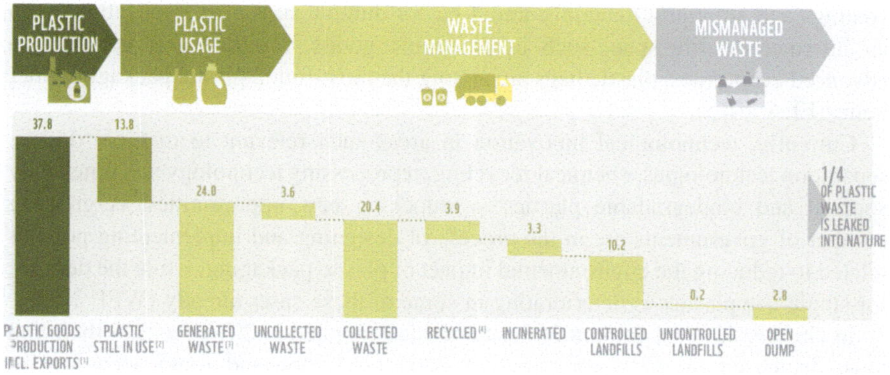

Fig. 9.1 Overview of the plastic lifecycle in the Mediterranean (million tonnes). (Source: WWF 2019)

Growing at 4% per year, plastic goods production in the Mediterranean reached almost 38 million tonnes in 2016. This represents 10% of all plastic goods produced globally, making the region the world's fourth largest plastic producer. The emissions by plastic production across all Mediterranean countries reaches approximately 194 million tonnes of carbon dioxide each year, accounting to approximately six times the annual carbon emissions of London (WWF 2019).

The generation of plastic litter in seas and oceans exhibits several environmental impacts. EFSA (2016) points out the presence of microplastics and nanoplastics in the food chain, with particular focus on seafood, raising concerns on the increased toxicity on the food products harvested at sea due to contaminants. Besides the direct impact in the quality of marine environment, marine littering has a significant socioeconomic impact. Marine littering threatens public health through the food chain which in turn reduces the catch of the local fishers and in turn contributes to loss of jobs (i.e. fishermen, tourism, etc.), property devaluation and population move.

The economic impact of marine litter is thought to be significant, especially when taking into consideration the above-mentioned health costs. Key economic sectors in the Mediterranean, especially fisheries and tourism, are negatively impacted by plastic pollution. Marine litter is estimated to cause an annual economic loss of €61.7 million to the EU fishing fleet (around 0.9% of annual total revenues) because of reduced catch and damage to vessels, while polluted beaches decrease tourist demand and consequently job losses in the touristic sector (UNEP 2016).

WWF (2019) Report supports that the main system failures resulting in plastic pollution across the entire plastic life cycle can be broken down in five stages: production; consumption; waste collection; waste treatment; and secondary markets for recycled material (Fig. 9.1). The lack of incentives for upstream innovation, tourism, lack of collection and recycling capacity and the low profitability complete the picture of main sources of pollution. Nonetheless, according to Plastics Europe (2015), the major source of waste is considered to be plastic packaging.

In Europe, plastic production comes in three broad categories: about 40% for single-use disposable applications, such as food packaging, agricultural films and disposable consumer items; 20% for long-lasting infrastructure such as pipes, cable coatings and structural materials; and 40% for durable consumer applications with an intermediate life span, such as electronic goods, furniture and vehicles. In advanced economies, plastic bags are among the most-found plastic packaging litter items (EEA 2015).

Currently, technological innovation in areas such relevant to material design, separation technologies, chemical recycling, reprocessing technology and renewably sourced and biodegradable plastics is unlocking new opportunities. A growing number of governments are in the process of designing and implementing policies related to reducing the environmental impact of plastic packaging, while the demand for single-use plastics is deteriorating in some of these cases already (WEF 2016).

In Greece, the total amount of plastic waste generated in 2016 was 257 thousand tons, decreased by 65% compared to 2006 (755 thousand tons) according to Eurostat. Almost 39 thousand tons of plastic waste were recycled, while only 1000 ton was landfilled. Greece appears in the second lowest position, after Cyprus, in terms of per capita volume of processed and recycled plastic waste (4 kg per person), according to Eurostat data. However, the relevant published data by the Hellenic Recycling Agency differ significantly, signifying underestimations from the side of Eurostat and the need for further and deeper research.

In addition, according to the recyclable raw materials trade data of Eurostat, about 61 thousand tonnes of recyclable plastics were exported to other countries (inside and outside the EU) (IOBE 2019).

The wider plastic industry is an important driver of growth for the Greek economy. The total contribution of the sector from its operation is estimated at €3 billion or 1.6% of the country's GDP in 2018. In terms of employment, total contribution is estimated at 67.2 thousand jobs (direct, indirect and induced impact), or 1.8% of total employment in the country, while public revenues from taxes and contributions due to the operation of the sector exceed €900 million (IOBE 2019).

The above environmental and socioeconomic characteristics depict the need for a transition to a new era where single-use plastics are eliminated and recycling rates of plastics is increased. In addition, elements that give added value and increase competitiveness of the industry globally are necessary. The transition of the Greek plastics industry should modernise the production process and proceed with adopting circular models in their production. This will not only strengthen their presence in the market but will also help the skills upgrade of the existing employment. In other words, the reform of the sector shouldn't only consider the environmental, but the social and economic aspects as well.

9.2 Policy Mapping on Plastic Reduction and Circular Economy

WEF (2016) reports that "The circular economy is gaining growing attention as a potential way for our society to increase prosperity, while reducing demands on finite raw materials and minimizing negative externalities. Such a transition requires a systemic approach, which entails moving beyond incremental improvements to the existing model as well as developing new collaboration mechanisms."

In January 2018, the European Commission launched the "EU Plastic Strategy" (European Commission 2020) to reduce single-use plastic pollution followed by the EU Directive on the "Reduction of the Impact of Certain Plastic Products on the Environment" (SUPD), published in June 2019 and entered into force in July 2019 to tackle the plastic waste issues in the European countries and reduce marine plastic litter. The SUPD requires Member States (MS) to prohibit certain SUPs items (cutlery, plates) use and requires MS to take the necessary measures to achieve an ambitious and sustained reduction in the consumption of several other SUPs products. It is worth noticing that the SUPD covers single-use plastic items including bio-based and biodegradable plastics regardless of whether they are derived from biomass or are intended to biodegrade over time as well as those made of different materials (multi-layered or composite materials), such as plastic-coated paper or plastic-lined cartons.

As explicitly stated in Article 2, coherently with the "waste hierarchy" within the Directive 852/2018 and the EU circular economy approach (EU action plan for the Circular Economy – EU COM/2015/0614 – and European Strategy for Plastics in a Circular Economy, EU COM/2018/028), the SUPD promotes circular approaches that give priority to sustainable and nontoxic reusable products and reuse systems rather than to single-use products, aiming first and foremost to reduce the quantity of waste generated. The restrictions for food packaging and beverage cups introduced by the SUPD offer the opportunity to scale up reusable alternatives rather than simply switching to other single-use-based material. However, this requires a systemic and widespread behavioural change to move away from single-use plastics, towards reusable products and systems. Despite several solutions being developed and applied locally to prevent SUPs products across European regions and cities, there is still a lack of policy support for the widespread of these alternative through incentives for innovation adoption of non-SUPs products to enhance systemic change and enable a drastic reduction of plastic consumption thus reduction of plastic production and CO_2 emissions.

For a long-term, environmental policy in Greece mainly focused on waste management (e.g. reduction and reusage), where CE projects were fragmented, often considered identical to material recycling. Following the EU legislation and the Communication of the Action Plan for the circular economy, the Hellenic Ministry of Environment and Energy adopted a National Circular Economy Strategy (NCES) in 2018 capturing a refined methodology of implementing CE in Greece (Ministry of Energy and Environment 2018). Although significantly delayed, the

NCES is an excellent list of topics to be discussed as basic themes for a future implementation plan. One of its main drawbacks is the fact that it neglects to address core challenges of the country. Specifically, lagging regions, suffering from persistently low private investments and limited bank liquidity, tend to adopt short-term, survival solutions. Thus, a prerequisite for the NCES to succeed is a detailed context-specific analysis of cooperation, coordination and synergies to come up with solutions shifting from a short-termism behaviour to a realistic, profitable, long-term strategy and the corresponding action plan (Koundouri et al. 2019).

9.3 BL.EU. Climate: Climate Innovation in Southern Waters

The main objective of the EIT Climate-KIC BL.EU. Climate project, that was implemented in 2019, was to address the challenge of plastic marine littering in southern European waters by building capacity for innovation to address the issue at the very beginning of its life cycle, on the prevention side and plastic waste reduction with significant climate change mitigation potential from the reduction in the collected and handled plastic waste. Greece, Portugal and Croatia gathered around this problem and identified three pillars around ports (commerce, fishing, tourism) working closely with local problem owners: in Croatia, islands Cres Zlarin; in Greece, the port of Piraeus, islands of Milos and Andros; and in Portugal, the port of Lisbon (BL.EU. Climate 2020).

The project started by conducting an extensive stakeholder mapping in all areas. Secondly, validation interviews/surveys were performed based on a common questionnaire designed by all project beneficiaries, which was targeting tourists of the above-mentioned regions. The results of the questionnaires were analysed and presented at different workshops conducted in the project sites. The main objective of the workshops was to trigger discussions among the participants (mostly stakeholders identified at the mapping exercise) on potential solutions to prevent, reduce and collect marine litter, focusing mostly on plastics. All the above led to the design of a strategic roadmap by all three countries, identifying steps to reduce the negative effects caused by plastic waste in the future, supporting not only governments but also regions, municipalities, industries, consumers and civil society to improve the awareness campaigns, systems design, replacement, refuse, recycling and reuse of plastic.

The outcomes of the BL.UE Climate project for the Greek case studies (Piraeus, Milos and Andros islands) are presented in the next section.

9.3.1 Stakeholders Mapping and Problem Identification

The project team chose four tools from the EIT Visual toolbox for system innovation, problem definition and stakeholder mapping in all project countries (De Vicente Lopez and Matti 2016). Below, the stakeholder mapping for the Greek case and the corresponding methodology is presented.

Nailing down the problem and identifying its different components and details has been the first objective of the research team, which aimed at arriving at a common ground for future actions towards the reduction of marine litter in the seas of South Europe. The use of the Pentagonal problem tool (Annex 9.1) identified the key dimensions related to this problem and facilitated the understanding of that complex and multidimensional challenge, which extends to climate change, societal, economic, resource and technical challenges. The common climate change challenge is linking the generation of marine litter with emissions from the collection and from sound waste management. Regarding the social and economic challenges for all locations, reducing littering especially during touristic seasons and managing the negative economic impact on tourism and fisheries emerge as priorities for all case studies, with non-existing recycling infrastructure on the islands of Milos and Andros being an additional challenge to overcome. Funding is another key aspect in terms of resources that drives the response velocity of the Greek islands in confronting the marine litter problem.

A significant number of stakeholders are involved in the plastic litter issue. To list and categorise the myriad of stakeholders around the project, the Actor tree canvas tool (Annex 9.2) was used. The common ground and the common challenge to develop a Roadmap for Plastic Free southern European waters led to the identification of 14 common stakeholder groups, from fishermen, tourists, citizens and start-ups related to maritime to policy makers, NGOs, researchers, universities and schools, foundations, shipping companies and plastic and recycling industries. Specifically, a key stakeholder in the island of Milos seems to be a raw material extracting company.

The building of a stakeholder profile was allowed by the Enlarged empathy map tool (Annex 9.3) by quickly browsing the sources of information available to any individual. Its great value lies in the delivery of a clear and accurate profile of the stakeholders. Mostly affected by the plastic pollution of the seas are the fishermen, while the biggest polluters seem to be the tourists and big companies, which benefit other key stakeholders (e.g. local maritime industry, SMEs). Finally, policy makers seem to play a key role in preventing or perpetuating the issue.

Lastly, in order to identify the interdependencies among the core players in the plastic industry, the Interest-Influence-Adaptation map tool (Annex 9.4) was used, which demonstrated where stakeholders stand when evaluated against the same key criteria and compared to each other. In all Greek case studies, policy makers are ranked medium to in all three key criteria (e.g. interest, influence and adaptation), being an undeniable key stakeholder in this challenge. Besides the high interest and adaptability to change of fishermen, they have limited influence on the issue, being, thus, dependent on more influential stakeholders, such as the municipality, or in the case of Andros the tourists.

The engagement strategy of the primary stakeholder identification analysis presented above will aim at the engagement of the stakeholders at different levels using a number of fora in order to integrate their input within the project. The three key stakeholder types, as identified in the analysis above, were interviewed in order to validate the presented outcomes, while tourists were interviewed in regard with their awareness and willingness to pay for the transition into higher quality and more sustainable ecosystems. Lastly, a participatory workshop, which took place at the end of the project, aimed at bringing together core actors related to the plastic pollution in these areas in order to exchange views on the potential solutions.

9.3.2 Survey and Conclusions

The analysis of the questionnaire responses from the three locations in Greece (Piraeus, Milos, Andros) highlighted findings that are relevant to the wider range of stakeholders related to marine litter, such as policy makers at national and regional levels, entrepreneurs of the tourist sector, tourists and researchers.

As a first finding, most of the respondents understand that the natural environment is at crisis, and this is caused mainly by anthropogenic factors. Human intervention is perceived to have had bad consequences at the environmental ecosystems, while most of the participants in the survey mentioned that humans do not have the right to modify the natural environment based on their own needs, which after all is one of the basic notions of environmental sustainability.

In contrast to the above, the survey revealed that even though a level of environmental awareness exists, in some cases this is considered to be superficial. Around one fourth of the participants do not understand core environmental problems and therefore cannot be engaged in further actions, either during their vacations (i.e. stop using single-use plastics at the beach) or in their home country. On top of that, most of the respondents mentioned that they are not aware of the Plastics Directive (European Parliament 2019). This finding indicates a gap in engaging in education and information of the public, while it is consistent with findings from the AdaptInGR project (LIFE-IP AdaptInGR 2019), which identified that 22% of Greek respondents were "a little" or "not at all" informed about climate change.

It was also evident that the participants could not identify the importance of sustainability and its potential links to environmental and economic systems. A striking example is that a significant number of people were not aware of the extent of the marine plastic pollution and its detrimental effects on the marine ecosystems.

That gap was evident across almost all questions asked. For example, many of the participants did not consider the natural environment as part of their own living niche. In fact, the analysis showed that they were willing to pay (WTP) more for a clean hotel room but less for a cleaner/greener ecosystem. It is obvious that despite acknowledging the importance of sound environmental management, these respondents could not understand how environmental degradation could affect them and therefore considered it as a problem related to the wider society. Most of them were not willing to pay more than €5 to receive green environmental services during their vacations.

In addition to the above, even though a significant part of the sample identified the importance of preserving the natural environment and have already stopped (or are willing to stop) using plastic bags, plastic straws and other single-use plastics (Fig. 9.2), few were aware of key European policies on plastics (such as the Plastics Directive). In more structural characteristics, higher levels of awareness were denoted among younger people, whereas people belonging to older age classes appear to be more reluctant in understanding core elements of sustainability and designing adaptation measures. Higher-income families of the sample have not yet developed the need to purchase greener vacations, a finding which is significantly different to the tourist sectors of other EU Member States (Italy, Sweden, Norway, etc.).

Even though a great share of the responses identified the significance of environmental responsibility, it was observed that the WTP was low. Additionally, when the cost factor was removed from their decision, the respondents were willing to adapt their needs. When the participants in the study were asked to rate different behavioural change strategies, they showed preference to methods that do not involve additional efforts or costs.

9.3.3 *Design of the Roadmap*

The roadmap is a strategic tool co-created by all partners in the three countries aiming at providing insight for all involved parties on needed measures to reduce plastic waste in the Mediterranean Sea. The 10-year roadmap developed for the BL. EU. Climate project identifies the actions that need to be taken by policy makers, the private sector and the society in the future in different time spots (2020, 2025, 2028, 2030, 2030 and beyond) so that the environmental footprint of single-use plastics is reduced drastically. This can happen by avoiding unnecessary packaging and banning non-recyclable plastic. These actions, if properly designed, can lead to modern business models that stimulate growth and jobs. By complying with the EU regulations, the Member States can support the shift of regulating the plastics industries to more environmental processes in the production line, to the production of higher added value materials (e.g. bioplastics, recyclable plastics) and to apply environmental standards in the R&D stage, i.e. in the design of the product.

Also, Member States could promote environmental and sustainable strategies according to EU regulations, by funding research projects in partnership with private companies (e.g. European structural and investment funds by European Commission), to identify the best available techniques for alternative materials to replace single-use plastics and support innovation, by providing necessary infrastructure (i.e. public fountains for safe drinking water) and by driving public debate and supporting youth delegations such as "Zero Plastic" ambassadors for international events.

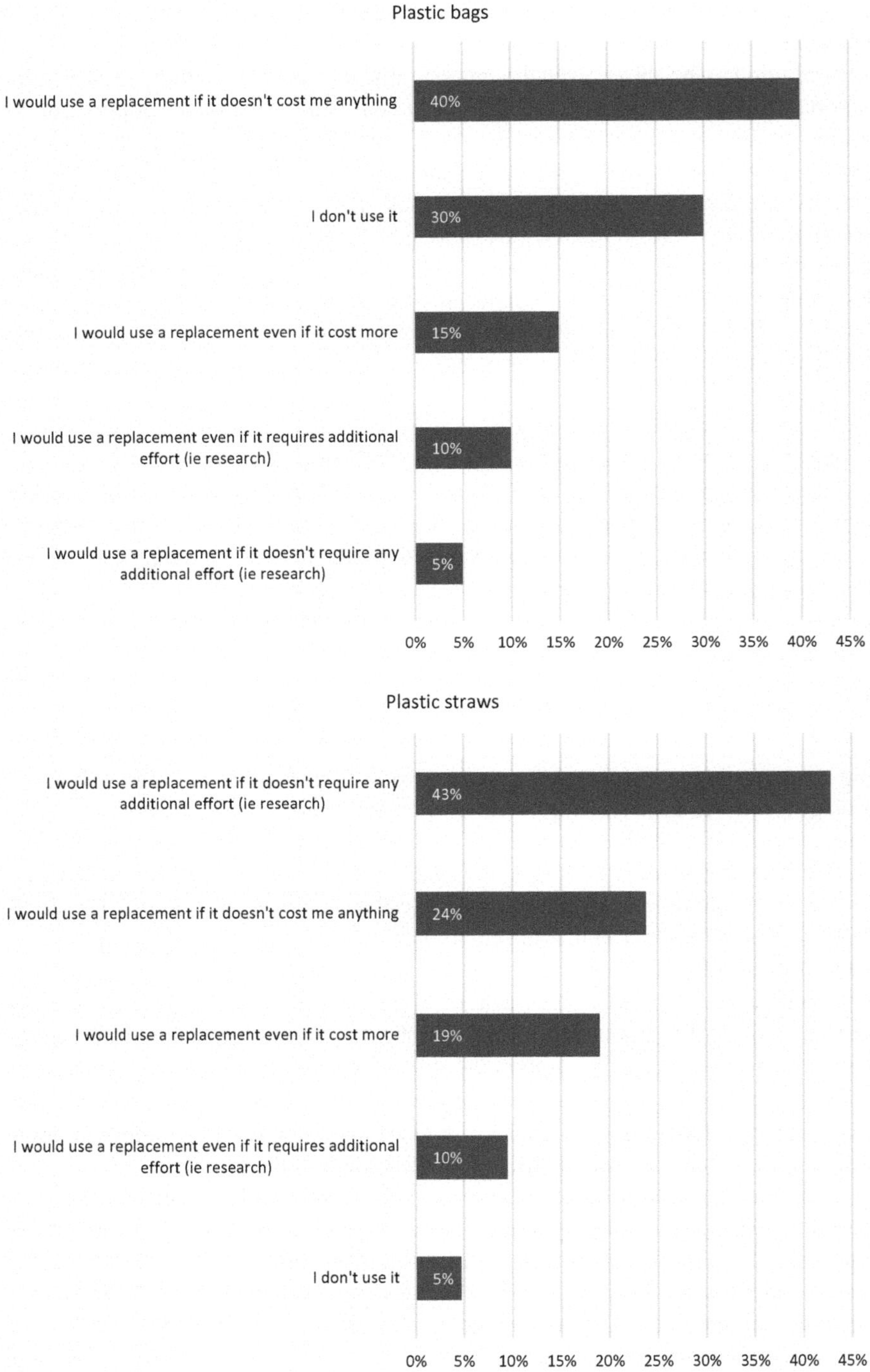

Fig. 9.2 Perceptions on the replacement of single-use plastics

Once again education is key to understand the environmental impact of plastic littering and how it affects public health. The educational system needs to be rethought/redesigned at the government level, leading schools and universities to avoid plastic, promoting discussion and change. Equally, education of policy makers and decision-makers is considered important. In a regional and European level, it's important to consider ourselves as an inseparable part of the natural ecosystem, taking up responsibility, embracing climate-friendly actions in order to prevent harmful effects to our seas and oceans. The Paris Agreement should be seen as a navigator to investment and financing opportunities around climate change, strengthening the role of the private sector in taking climate action.

The private sector should invest on new business models that integrate not only the economic but the social and environmental standards required by the society and the government. Models such as circular economy (CE), sharing economy and blue economy are already becoming reality, but there is a need to expand and apply those models on traditional sectors such as food industry. The private sector could be a pioneer and increase competition by providing guidelines for tourists and encouraging good environmental standards, such as zero waste and recycling practices. Private companies could also be involved in deposit-return schemes, acting against the throw-away culture and banning the multi-layered packaging that cannot be recycled. Also, the private sector needs to play a key role in developing packaging design, materials and technologies in line with the circular economy, which provide sustainable solutions to valorise used plastics and thus reduce plastic waste (Ellen MacArthur Foundation 2017).

The civil society composed of non-governmental organisations and youth-led movements can create awareness campaigns that target different audiences at different levels. Civil society organisations can lead plastic-free events in different scales, at neighbourhood level, or plastic-free campus on universities, creating partnerships with private sector pursuing to pressure the governments to provide funding and support investments towards plastic waste elimination. It is important to create communication for high-level politicians and industry as well. Non-governmental or civil sector organisations should lead and help establish a platform for sharing information and supporting change makers. It is necessary to engage communities in order to increase citizen-science and promote recycling methods and informational campaigns (various targeted groups, from schools to popular tourism destinations).

9.4 MEDfreeSUP Project: Tacking Single-Use-Plastic Item Uses in the Easter Mediterranean Sea

The EIT Climate-KIC MEDfreeSUP project, kicking-off in 2020 for a 2-year period, falls in the plastic waste prevention (PWP) approach developed by EIT Climate-KIC eCircular programme, to enable local ecosystems to move towards reusable materials. This is foreseen as one of the most efficient solutions for addressing the single-use plastics packaging problem, providing tangible economic, environmental and social benefits. The project will focus on the East Mediterranean coast, targeting the three biggest coastal countries: Italy, Croatia and Greece, with pilot sites in the Italian region of Emilia-Romagna (Bologna, Cervia, Misano Adriatico, Ravenna), Aegean Sea (Greek islands of Syros and Ikaria) and Ionian Sea (Greek island of Corfu and Croatian islands of Zlarin and Cres).

The main objective of the project is to set a replicable voluntary protocol for free single-use plastics items related to food packaging in cafes, restaurants, foods stores, hotel, beach facilities as well as for public events and public spaces. The purpose of the protocol, to be designed as a toolbox, is to provide support and guidance to local businesses in order to comply to the newly EU SUP Directive[2] (2019) and go beyond the law to engage Mediterranean islands and cities in the transition towards a free single-use plastic environment. The project will follow a system innovation approach, benefiting from EIT Climate-KIC extensive network of innovative start-ups and research facilities which can support the transition towards single-use-plastic (SUP) free environment. The involvement of problem's owners (local stakeholders with the need of eliminating single-use-plastic items) will play a central role in the project, being at the core of the strategy to develop the protocol. The project will create a learning loop between the scientific community, the industry, policy makers and local businesses in order to identify effective alternative solutions to the use of SUP items, thus limiting the consumption of plastic products which will in turn decrease the production of plastic item itself and plastic wastes. Therefore the project ought to have positive impacts both at the start and at the end of the plastic chain, reducing CO_2 emissions and marine plastic litters.

The project will build-up on pre-existing projects focusing on marine plastic litter and waste management (i.e. Interred Mediterranean ACT4LITTER and Plastic Busters projects) and work closely with local stakeholders and business owners to assess their needs and co-identify alternative solutions and new innovative approaches to replace SUP items. The inventory of potential solutions for free SUP item in bars, cafes or restaurants will include a legislative, environmental and financial assessment in each country to ensure that the proposed alternatives are legally applicable, environmentally and financially sustainable to avoid creating new negative environmental externalities and financial burdens. Additionally, through the protocol, MEDfreeSUP seeks to assess the capacity of public authorities (from local to national and EU level) to support local businesses in the transition to the use of non-SUP items and ensure the smooth implementation of the EU SUP Directive. In the current Covid-19 context, the financial capacity of local business owners in coastal touristic areas (main targets in Greece and Croatia pilot sites) to support the potential additional costs and changes in the ways the business should operate when it comes to using alternative products to plastic items is even more relevant if we

[2]Directive (EU) 2019/904 of the European Parliament and of the Council of 5 June 2019 on the reduction of the impact of certain plastic products on the environment

wish local stakeholders to voluntary accept the switch to a new environment-friendly business model without SUP items.

To sum up, the protocol will provide a guideline for local business to eliminate SUP items by presenting the alternatives available at country level and proposing supporting tools for the implementation of these alternatives, in cooperation with relevant public authorities and business service providers. The MEDfreeSUP project will additionally tackle behavioural change in consumers (tourists and local residents) by setting an online awareness platform following a gamification approach. It will engage consumers in being proactive in their choice of consumption with a rewarding skim to stimulate the consumers' behavioural change and adoption of these solutions. The protocol and online platform will be implemented and tested in voluntary pilot cities and islands to prove their effectiveness and viability and ensure their replicability in other European regions.

9.5 Conclusions

The present chapter refers to the issue of plastic marine litter in the wider Mediterranean region and especially in Greece. Plastics account for 95% of the waste in the open sea, on the seabed and beaches across the Mediterranean, placing a burden on the environmental, social and economic sustainability of the area. In January 2018, the European Commission launched the "EU Plastic Strategy" to reduce single-use plastic pollution followed by the EU Directive on the "Reduction of the Impact of Certain Plastic Products on the Environment". In Greece, waste management remains one of the most pertinent environmental problems. More than 80% of the generated waste are landfilled, while recycling accounts to 20%. Inconsistencies were noted in the data regarding plastic waste generation and management between official authorities depicting the importance of further research into this.

To mitigate the impacts of plastic litters in the Mediterranean, EIT Climate-KIC funded two projects, BL.EU. Climate (implementation period: 2019) and MEDfreeSUP (implementation period: 2020–2021), on creating knowledge and supporting the plastic reduction in the Mediterranean seas. The BL.EU. Climate project addressed the challenge of plastic marine littering in southern European waters by building capacity in three Mediterranean countries: Greece, Portugal and Croatia. The MEDfreeSUP project, which is the continuation of the BL.EU. Climate project, aims to set a replicable voluntary protocol for free single-use plastics food packaging adoption for the private commercial sector and for public places in three Mediterranean countries: Greece, Italy and Croatia to provide support and guidance to local business in complying with EU Directives.

According to BL.EU. Climate results, one of the greatest barriers in Greece is the knowledge gap across stakeholders. Most of the respondents understand that the natural environment is at crisis and this is caused mainly by anthropogenic factors. However, the most affected by marine litter actors (e.g. fishermen and tourists) seem to lack a basic understanding of the marine plastic pollution and its detrimental effects on the marine ecosystems, as well as the existence of European Directives on plastics. Fishermen specifically seem to know about the marine litter problem, but they ignore the impacts on their professional and personal life, especially about the microplastics and how they end up in the food chain. Another outcome of the study was that besides the environmental responsibility being perceived as significant by the majority of tourists, their WTP for environment-friendly services was comparatively lower to their WTP for a renting better facility. However, when the cost factor was removed from the decision-making choice, the respondents were willing to adapt their needs. Lastly, elderly people are observed to be less supportive of the environmental cause, despite identifying the importance of having good environmental status.

Finally, the project produced a 10-year roadmap identifying the actions that are needed to be taken by policymakers, the private sector and the society in the future in different time spots to tackle the plastics issue. Top priorities appear to be education and private initiatives in combination with regulation and policy implementation (e.g. circular economy strategy). Education is key to comprehend the damages of plastic and how it affects public health, while the awareness raise could bridge the knowledge gap of adults, who have completed their secondary education. The private sector can be a pioneer in driving change by implementing circular economy models and creating opportunities supported by complementary laws and strategies.

Annexures

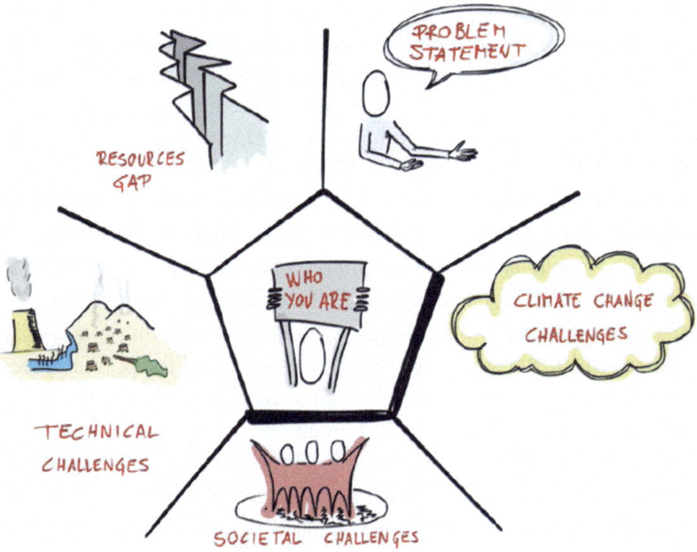

Annex 9.1 Pentagonal problem tool

Annex 9.2 Actor tree tool

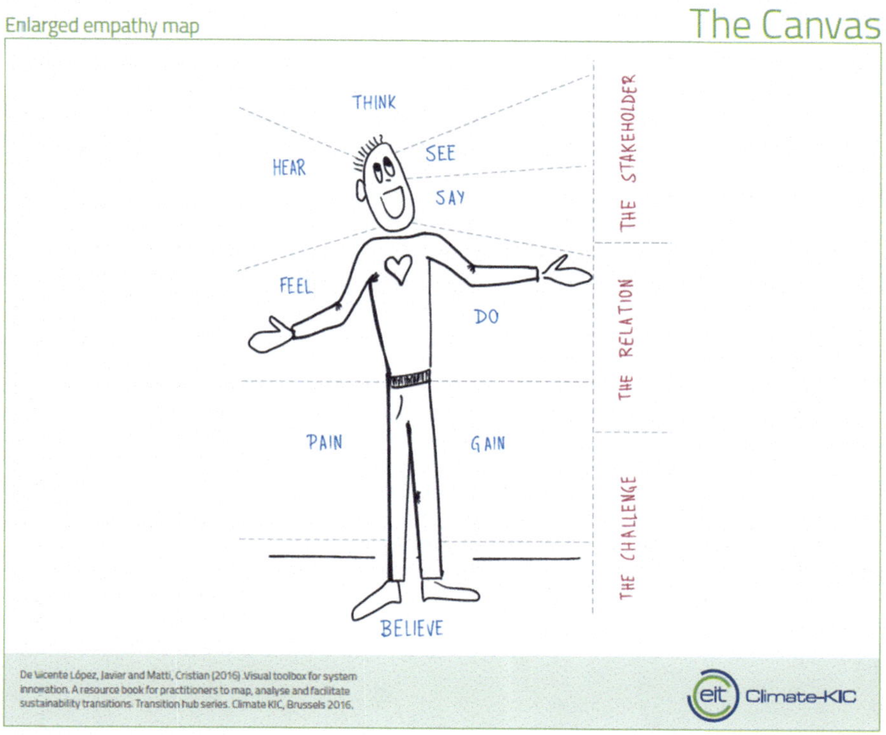

Annex 9.3 Enlarged empathy map

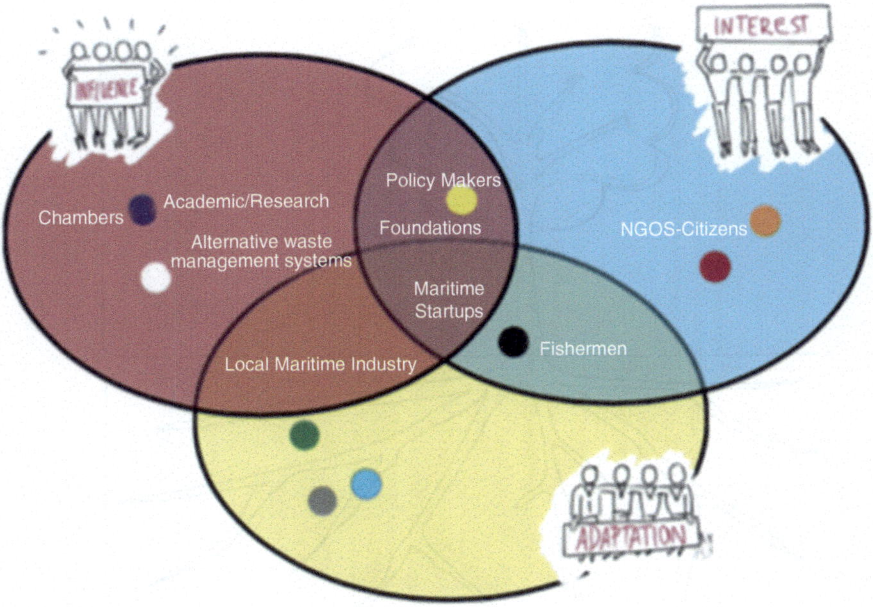

Annex 9.4 Interest, influence and adaptation map

References

BL.EU. Climate. (2020). *EIT climate-KIC project. Climate innovation in Southern waters*. https://www.athenarc.gr/el/climate-innovation-southern-european-waters-bleu-climate

De Vicente Lopez, J., & Matti, C. (2016). *Visual toolbox for system innovation. A resource book for practitioners to map, analyze and facilitate sustainability transitions* (Transitions Hub Series). Brussels: Climate-KIC.

EEA. (2015). *Marine Litter study to support the establishment of an initial quantitative headline reduction target – SFRA0025; European Environment Agency, Top marine litter items on the beach*. http://www.eea.europa.eu/data-and-maps/daviz/marine-litter-items-on-the-beach

EFSA. (2016). Presence of microplastics and nanoplastics in food, with particular focus on seafood. *EFSA Journal*. https://efsa.onlinelibrary.wiley.com/doi/full/10.2903/j.efsa.2016.4501.

EIT Climate-KIC. (2020). *Europe's leading climate innovation initiative*. https://www.climate-kic.org/who-we-are/what-is-climate-kic/

Ellen MacArthur Foundation. (2017). *The new plastics economy: Rethinking the future of plastics*. http://www.ellenmacarthurfoundation.org/publications. Accessed 19 Dec 2019.

European Commission. (2020). *European strategy for plastics in a circular economy*. https://ec.europa.eu/environment/waste/plastic_waste.htm. Accessed 10 Apr 2020.

European Parliament. (2019). *Directive (EU) 2019/904 of the European Parliament and of the Council of 5 June 2019 on the reduction of the impact of certain plastic products on the environment*. https://eur-lex.europa.eu/eli/dir/2019/904/oj

IOBE. (2019). *Foundation for economic and industrial research. The plastic sector and its contribution to the Greek economy*. http://iobe.gr/docs/research/RES_05_F_10122019_REP_GR.pdf

Koundouri, P., Tsipouri, L., Papadaki, L., & Argirou, M. (2019). *The Greek circular economy approach: Mapping circular economy policies, instrument, initiative & action plans in the public sector*. http://wpa.deos.aueb.gr/wpa_show_paper.php?handle=1907

LIFE-IP AdaptInGR. (2019). *LIFE-IP AdaptInGR – Survey for climate change*. Athens: LIFE-IP AdaptInGR.

Ministry of Energy and Environment. (2018). *National circular economy strategy*. http://www.greekscrapmetal.gr/assets/uploads/files/ethniki_stratigiki_kikliki_oikonomia.pdf. Accessed on 10 Sept 2019.

PlasticsEurope. (2015). *Plastics – The facts 2015*. Retrieved from Brussels: http://www.plasticseurope.org/Document/plastics-the-facts-2015.aspx

UNEP. (2016). *Marine plastic debris and microplastics: Global lessons and research to inspire action and guide policy change.*.

UNEPMAP. (2020). *Mediterranean 2017 quality status report*. https://www.medqsr.org. Accessed on 5 April 2020.

UNWTO. (2014). *UNWTO tourism highlights*, 2014 Edition. eISBN: 978-92-844-1622-6. https://www.e-unwto.org/doi/book/10.18111/9789284416226

WEF. (2016). *The new plastics economy. Rethinking the future of plastics*. http://www3.weforum.org/docs/WEF_The_New_Plastics_Economy.pdf

WWF. (2018). *Report. Out of the plastic trap: Saving the Mediterranean from plastic pollution*. http://awsassets.panda.org/downloads/a4_plastics_med_web.pdf. Accessed on 10 April 2020.

WWF. (2019). *Report. Stop the flood of plastics: How Mediterranean countries can save their sea*. https://www.wwf.fr/sites/default/files/doc-2019-06/20190607_Rapport_Stoppons_le_torrent_de_plastique_WWF-min.pdf. Accessed on 5 May 2020.

Chapter 10
Sustainable Shipping: Levers of Change

Andreas Papandreou, Phoebe Koundouri, and Lydia Papadaki

Abstract Sustainable shipping refers to the broad set of challenges, nature of governance rules and regulations, patterns of management and corporate behaviors and aims, engagement of stakeholders, and forms of industrial activity that should come to define a marine transport industry that is shaped by the broader societal goals of sustainable development. This chapter aims to provide a brief overview of the marine transport industry, its role and relevance in sustainable development, and the kinds of changes that are needed for shipping to be sustainable. The focus is mostly on the environmental dimension of sustainable development. As a sector, and for reasons that have to do with the special nature of its international governance that partly falls outside the confines of national jurisdictions, shipping may have been a late comer to some of the most pressing sustainability challenges of our time. After presenting some recent economic trends of the sector and their potential implications for sustainability, the chapter will present some environmental pressures that are related to shipping and will focus on two particular sustainability challenges confronted by maritime transport: the need to drastically reduce sulfur emissions and the even more demanding challenge to mitigate CO_2 emissions. Before concluding, the penultimate section will briefly present some sustainability initiatives already under way.

A. Papandreou (✉)
Department of Economics, National and Kapodistrian University of Athens, Athens, Greece

United Nations Sustainable Development Solutions Network Greece, Athens, Greece
e-mail: aap@econ.uoa.gr

P. Koundouri
School of Economics and ReSEES Laboratory, Athens University of Economics and Business, Athens, Greece

United Nations Sustainable Development Solutions Network Europe, Paris, France

EIT Climate–KIC Hub Greece, ATHENA Research and Innovation Center, Athens, Greece

ICRE8: International Centre for Research on the Environment and the Economy, Athens, Greece

L. Papadaki
United Nations Sustainable Development Solutions Network Greece, Athens, Greece

EIT Climate–KIC Hub Greece, ATHENA Research and Innovation Center, Athens, Greece

P. Koundouri (ed.), *The Ocean of Tomorrow*, Environment & Policy 57,
https://doi.org/10.1007/978-3-030-56847-4_10

Keywords Sustainable shipping · Maritime transport · CO_2 emissions mitigation · EU ETS

10.1 Introduction

Two landmark agreements adopted in 2015 are the 2030 Agenda for Sustainable Development under the framework of the United Nations Sustainable Development Summit and the Paris Agreement on climate change under the auspices of the United Nations Framework Convention on Climate Change (UNFCCC). None of the 17 sustainable development goals (SDGs) is dedicated to the thematic area of transport. In elaborating the goals, the international community recognized that by integrating and mainstreaming transport considerations into a range of SDGs, its cross-sectoral nature would be a critical enabler of most of them (Benamara et al. 2019).[1]

Maritime transport is an economic sector in its own right. With 80% of international merchandise trade by volume and over two thirds by value in 2017 (UNCTAD 2018), it is central to the sustainability agenda. Maritime transport links almost all countries' relevant supply chains, supports international production processes, carries international trade, and provides access to the global markets. In addition, many sectors and industries are intimately linked to marine transport: marine equipment manufacturing, marine auxiliary services (e.g., insurance, banking, brokering), fisheries, tourism, ship building and demolition, and offshore energy (Benamara et al. 2019).

Maritime transport can be seen as environmentally friendly relative to other modes of transportation when measured in ton-miles (weight per distance traveled). In conjunction with its strategic economic and social function of supporting international trade, it can be viewed as an important sustainable development enabler (Benamara et al. 2019). Unsustainable transport patterns, however, are linked to numerous social costs in the form of air and marine pollution, GHG emissions, resource depletion, and biodiversity loss, among others.

Sustainability in maritime transport involves, inter alia, the ability to provide transportation infrastructure and services that also further the multiple dimensions of sustainable development, for instance, safety, accessibility, social inclusivity, reliability, fuel efficiency, affordability, clean environment, reductions in GHGs, and climate resilience. Figure 10.1 provides an overview of the intersection between the three pillars of sustainable development as they relate to the marine transport sector.

[1]This chapter draws heavily on Benamara et al. (2019) and UNCTAD (2018).

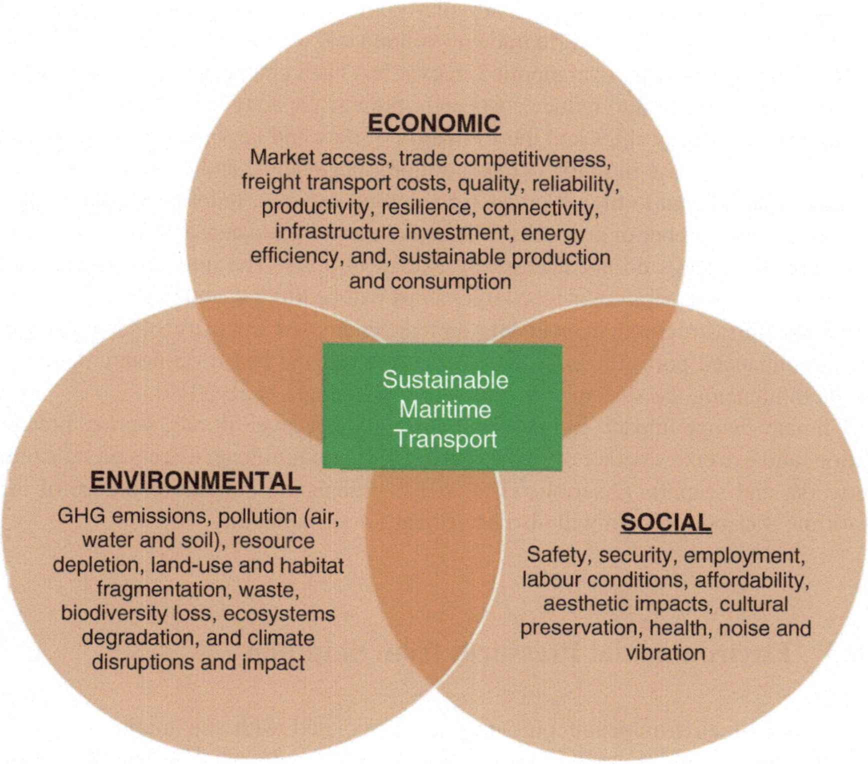

Fig. 10.1 Sustainable Maritime Transport. Source: UNCTAD (2015)

10.2 Recent Economic Trends for Maritime Transport

Demand for maritime transport increases in tandem with gross domestic product and industrial production. OECD (2017b) projects the tripling of total freight transport demand over the 2015–2050 driven mostly by economic growth with maritime transport accounting for 75% (up from 71% in 2015). The projected increase in total freight transport is expected to translate into 120% increase in CO_2 emissions (OECD 2017b).

The growing role of developing countries in trade and global economic growth is also reflected in marine transport with 60% of world seaborne trade volumes originating in developing countries, while 63% were delivered to their territories (UNCTAD 2018). New patterns of geographical distribution of production and consumption emerge altering cargo flows and directions with implications for shipping networks, fuel consumption, transport costs, ship emissions, and climate change (Benamara et al. 2019).

While there has been growth in deadweight tonnage of the commercial shipping fleet in the last two years, this has followed a 5-year period of decelerating growth.

The overall weak global demand particularly affected the container shipping segment which carries 16% of world trade by volume and over half by value (UNCTAD 2018). This period of an oversupplied market has been characterized by consolidation and rationalization to reduce costs and optimize capacity utilization as evinced in the arrival of mega ships and the formation of new and larger shipping alliances. The potential increase in market concentration, mega ships, and enhanced network efficiency could lead to higher prices for shipping services, redefine supply chains, and reduce the number of port calls. The economies of scale at sea brought about by large container ships do not necessarily extend to ports. The number of ports and terminals able to accommodate the larger ships will be limited, and ports will have to undertake infrastructural investments and the increased intensity of activity will require enhanced port efficiency. These are some of the broad economic developments influencing the sustainability equation (Benamara et al. 2019).

Climate change impacts in the form of rising water level, floods, storms, precipitation, and extreme weather events are likely to have significant effects on transport networks and seaports (Asariotis et al. 2017). Enhancing climate resilience of the maritime transport system will also be critical for sustainability.

10.3 Environmental Pressures from Shipping

A number of environmental pressures are associated with the marine transport industry. South and East Asia undertake a large share of ship recycling, and unsustainable conditions pose serious risks to human health, environment, and society including children. Hazardous and oily materials (e.g., asbestos, polychlorinated biphenyls, oils, and oil sludge) contained in many of the old ships are a key problem. A number of regulations are meant to address these issues (Benamara et al. 2019). Sea also Mikelis (2019) for fuller discussion.

Various types of wastes are generated by ships such as oily wastes, drainage from bilges, sewage and garbage, and cargo residues. Wastes dumped in the marine environment results in negative impacts in the form of chemical pollution and nondegradable waste that affects marine life while also degrading the natural and economic value of coastal areas. The MARPOL Convention addresses many of these concerns, and these obligations are mirrored in EU directives.

Harmful aquatic organisms and pathogens can be transferred between marine ecosystems through ships' ballast waters and sediments. This is a major environmental challenge that can significantly damage coastal and marine environments and ecosystems (Benamara et al. 2019). In September 2017 the International Convention for the Control and Management of Ships' Ballast Waters and Sediments came into force, requiring ships to have ballast water treatment systems.

About half of global crude oil production is carried by sea making oil spills a major pollution risk. The international oil pollution regulatory framework under IMO has contributed to a substantial drop in the number of oil spills from tankers.

Other types of pollution, including spills of hazardous and noxious substances, remain a concern.

10.4 New Challenges to Sustainable Shipping

10.4.1 Heavy Reliance on Oil for Propulsion

The transport sector accounts for more than 50% of oil demand today. Over half of the increase in freight transport energy projected by 2040 can be attributed to shipping (EIA 2017). Figure 10.2 provides a breakdown of energy usage in the transport sector globally in 2015. International shipping increased demand for energy at an annual rate of 1.6% from 2000–2014 (IMO 2014). Some decoupling between maritime transport activity and marine bunker fuel that has taken place in the recent past more likely reflects the upgrading of the global container fleet to larger more efficient ships, the scrapping of older ships and slow steaming rather than energy efficiency improvements or reduced dependence on oil. These trends were likely a response to the excess capacity resulting from the 2009 downturn (Benamara et al. 2019).

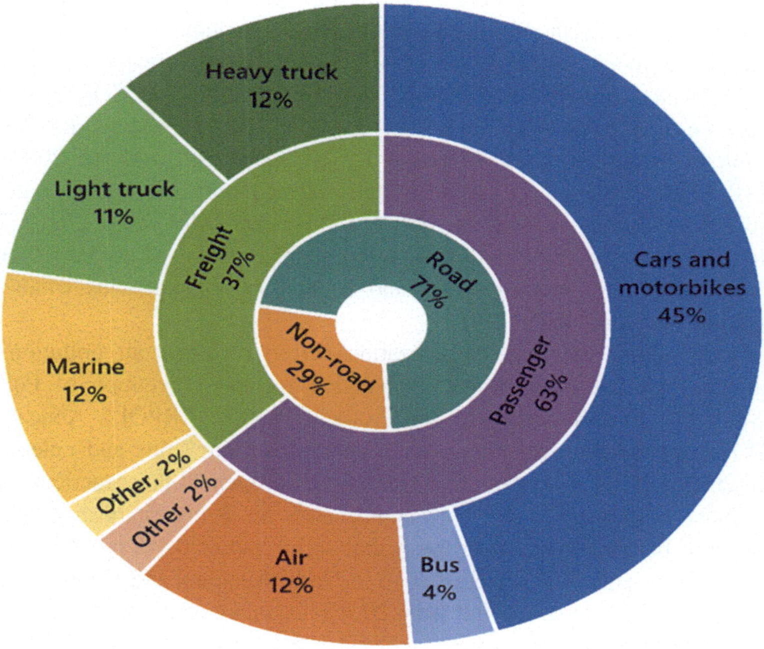

Fig. 10.2 Breakdown of energy usage in the transport sector globally in 2015. The outer ring gives the share of individual modes; the middle and inner rings aggregate these uses. (Data source: EIA (2016). Source: Bouman et al. (2017))

Reducing energy consumption and the heavy reliance on oil for propulsion is a key challenge for sustainable maritime transportation. Marine bunker fuels are very polluting and have a high carbon intensity. In addition, the affordability of maritime transport service could be jeopardized by high and volatile oil prices. A future of low oil prices, however, could undermine the needed transition sustainability.

10.4.2 Air Pollution with a Focus on the 2020 Sulfur Cap

Sulfur dioxide is not considered a greenhouse gas, and it has been argued that SO_2 has a cooling effect. The long-recognized harmful effects of sulfuric acid deposition to land-based ecosystems have led to strict regulation on land-based sources of SO_2 (power stations and vehicle emissions) with concomitant reductions of the sulfur burden over North America and Europe. Fossil fuel combustion from power plants is the largest source of SO_2 emissions (73%). Industrial facilities are the second largest source (20%) followed by smaller sources that include burning of high-sulfur-containing fuels by locomotives, non-road equipment, and large ships (EPA 2013). Emissions from shipping have been poorly regulated despite evidence of enhanced acidification in coastal regions. As a result shipping is responsible for a large proportion of man-made SO_2 emissions (Endres et al. 2018). In the year 2000, SO_2 emissions from shipping were three times greater than that from all traffic and aviation combined (Eyring et al. 2005). On some calculations a single container carrier emits as much SOx as 50 million of diesel cars (International Gas Union 2017). Shipping is thus by far one of the world's top sources of SOx as well as a major source of NOx and GHG emissions. NOx and SOx emissions from international ships account for about 13% and 12%, respectively, over the 2007–2012 period (IMO 2014). One large container ship visiting a port is estimated to produce the equivalent amount of NOx as that of 12,500 cars (McKinnon 2016). These emissions are closely associated with heavy bunker fuels. Ships have been an important cause of premature deaths and respiratory symptoms as winds carry marine emission inland (EPA 2016).

The International Maritime Organization (IMO) regulates air pollution from shipping through the International Convention for the Prevention of Pollution from Ships (1973) as modified by the Protocol 1978 (MARPOL). Annex VI of MARPOL specifically regulates airborne emissions from ships and entered into force in 2005. Emission control areas (ECAs) in Europe and the Americas enforce stricter limits on SOx emissions. New control areas are being established in China. Shipping will be required to meet the global sulfur cap of 0.5% by 2020. Stricter NOx ECA limits came into effect in 2016 in North America, and Northern Europe will apply NOx ECA to ships built from 2021 (Benamara et al. 2019).

The new IMO regulations due to take effect in January 2020 aim to drastically lower the sulfur cap for air emission from ships. The three main options available to shippers on current technology are to run on liquefied natural gas (LNG), to use HSFO and process air emissions through an exhaust gas cleaning system (EGCS) called "scrubber" fitted on board the ship along with dedicated tanks to hold and treat resulting wastewater from the process, or to switch from HSFO to a lower-sulfur fuel known as low-sulfur fuel oil (LSFO) (Halff et al. 2019). Many industry participants have yet to decide which of these three paths to take despite the imminence of the regulation and the fact that IMO first announced the cleaner-burning bunker rules as far back as 2008 (Halff et al. 2019).

Other fuel options include electricity, biodiesel, methanol, liquefied petroleum gas (LPG), Ethan dimethyl ether (DME), biogas, synthetic fuels, hydrogen, and nuclear fuel (GL 2017). The availability of fuels and their costs, uncertainty relating to alternative technologies and their level of maturation, and the investment requirements in terms of bunkering infrastructure are just some of the factors influencing the decision to adopt a particular option (GL 2017). Scrubbers, for instance, will require additional expenditures, and there are uncertainties about the underlying technology. Distillates are technically feasible, but if demand increases for them, the cost differential with conventional bunker fuels may widen. LNG use will involve important investments in bunkering facilities.

Halff et al. (2019) identify three sets of factors that discourage a prompt response to the new policies: the financial burden of premature compliance, financial risks stemming from market uncertainty (exacerbated by the IMO policy itself), and regulatory uncertainty. LNG and scrubber options both entail multimillion dollar up front capital expenditures. The attractiveness of the different options depends on the premium that low-sulfur fuel has relative to HSFO. The industry's rate of adoption of various compliance options will also likely affect their competitiveness, e.g., widespread adoption of LNG could lead to an increase in its price undermining its attractiveness. HSFO prices may plummet if there is large-scale switching to LSFO or LNG making scrubber adopters less competitive. In this sense, early adopters may suffer from a deficit of information (on what others are planning) providing an incentive for waiting.

Potential interaction with regulations on GHGs and NOx further adds to the uncertainty. A global cap on NOx or GHG could damage the business case for scrubbers that do not filter out NOx and are relatively carbon intensive. While LNG is low in both SOx and NOx, there are concerns that methane leakage could make it a source of high GHG emissions if the entire LNG lifecycle is accounted for. Similar concerns apply to LSFO or MGO being too carbon intensive if the fuel lifecycle of these fuels is considered (Halff et al. 2019). See Halff et al. (2019) for an extended discussion of these uncertainties.

10.4.3 SO_2 Emissions Policy

IMO has chosen performance standards ("obligation of results") over technical standards ("obligation of conduct") in order to regulate SO_2 emissions. It sets the amount of sulfur dioxide that ships are allowed to release in the air but leaves it up to the shippers to find the means (conduct) to achieve that goal. From a standard theoretical economic perspective that does not take into account transaction or enforcement costs, this makes sense. Performance standards allow participants to find the least-cost method of reducing emissions, whereas technical standards essentially "pick a winner" and can thus potentially stymie innovation and narrow the range of possible means of reducing emissions. Performance standards, however, are generally much more difficult to enforce. This is a special challenge in the case of shipping emissions as there is no single entity to carry out inspections on the high seas. Port states and flag states don't have the capacity and may not have the will to carry out inspections. New technologies, like remote sensing via satellites, may offer hope for effective enforcement (Halff et al. 2019).

10.4.4 CO_2 Emissions

All transports accounted for 24% of the world CO_2 emissions from fuel combustion in 2015 (OECD 2017a). Total shipping emissions reached approximately 938 million tons CO_2 emissions in 2012 with international shipping representing 85% of this total. Shipping accountsed for 2.2% of global total CO_2 emissions. Depending on economic growth and global energy demand, international carbon emission could increase by 50–250% by 2050 (IMO 2014).

International shipping emissions were notably absent from the Paris Agreement. CO_2 emissions from international shipping have grown more slowly than international trade. This decoupling reflects increases in shipping efficiency (with slow steaming, increased size of ships and other operational measures playing a key role rather than technological innovations). There is presently no global mechanism to control CO_2 emissions beyond the efficiency standards for new-build ships (Traut et al. 2018). The Kyoto Protocol mandated its parties to work through the IMO for emission reductions from international shipping. For international aviation emissions, it mandated the International Civil Aviation Organization (ICAO) (UNFCCC 1997). Parts of the shipping industry have argued that shipping should have a more limited role in emission reductions because of its "vital role" in serving developing economies (drawing on the notion of Common but Differentiated Responsibilities and Respective Capabilities) and because shipping has fewer opportunities to decarbonize relative to other sectors (ICS 2016).

IMO adopted a mandatory data collection system for fuel consumption of ship in 2016, and in April 2018 the IMO Marine Environment Protection Committee (MEPC) adopted an initial strategy on GHG emissions reductions from ships (IMO 2018). This strategy entails the first global climate framework for shipping and includes quantitative GHG reduction targets through 2050 as well as a list of candidate policy measures to help achieve these targets. A key target is to reduce CO_2 emissions per transport work as an average across international shipping by at least 50% by 2050 compared to 2008 while simultaneously pursuing efforts to a total phase out. Market-based measures (MBMs) are considered as potential measures. More generally the international community under the auspices of IMO/UNFCCC has seen a number of proposals in the form of incentivizing shipping companies to reduce carbon through operational changes or adoption of more carbon-efficient vessels, the introduction of a carbon tax on shipping, or emission trading mechanisms.

In the short term, CO_2 intensity of shipping can be reduced by a number of measures like changes to speed, ship size and utilization, retrofit technologies, and other efficiency measures. Slow steaming, a practice of deliberately lowering the speed of a ship to reduce fuel costs, is one suggested response to the sulfur cap. It proved very effective when the shipping industry was hit hard by the oil rally of 2002–2008. Slow steaming even in a lower oil-price environment can help mop up excess capacity when the shipping markets are oversupplied. In addition to saving energy, it has been argued that shipping carbon emissions are also reduced and marine transport reliability is improved by reductions in bottlenecks in terminals (Halff et al. 2019). See also Maloni et al. (2013) for a cost-benefit analysis of slow steaming.

Energy efficiency is also an important means of reducing air pollution. One study that considered 22 potential ship efficiency measures found that a reduction of 33% of CO_2 emissions could be achieved by 2020 (ICCT 2011). Another study found that energy-saving could reduce CO_2 emissions by 50% by 2030 (Alvik et al. 2010). Energy efficiency has been promoted in the maritime transport sector through regulatory measures in force since 2013: IMO's Energy Efficiency Design Index (EEDI), Energy Efficiency Operational Indicator (EEOI), and Ship Energy Efficiency Management Plan (SEEMP) (IMO 2017).

Virtually full decarbonization will be needed in the longer term that will mean fleet-wide deployment of near-zero carbon ships. This is a great challenge given the very short time frame (Traut et al. 2018). Bouman et al. (2017) review around 150 studies to provide a comprehensive overview of CO_2 emissions reduction potentials and measures published in the literature and find that emissions can be reduced by more than 75% based on current technologies (and through a combination of the proposed measures) by 2050. See Fig. 10.3 as a snapshot of CO_2 emissions reduction measures and their potential impact. Also, for a marginal abatement cost (MAC) curve that presents the average marginal cost associated with alternative individual measures in CO_2 emissions reduction, see Fig. 10.4.

Psaraftis and Zachariadis (2019) highlight some issue in the discussion about the use of alternative fuels for marine use for GHG reductions. Many of what are called "clean burning" fuels may be correctly labeled as such when focusing on SOx, NOx, and particulate matter but not when the GHG footprint is considered. When considering the life cycle GHG footprint of nearly all proposed alternative fuels, they are worse than conventional liquid fuels (marine gas oil (MGO), marine diesel oil (MDO), or desulfurizer fuel oil). For instance, when taking into account its life cycle, methane slip LNG's global warming effect is much worse than conventional liquid fuels and possibly even worse than coal. See Psaraftis and Zachariadis (2019) for a discussion of the alternative fuels: natural gas (NG), liquified natural gas

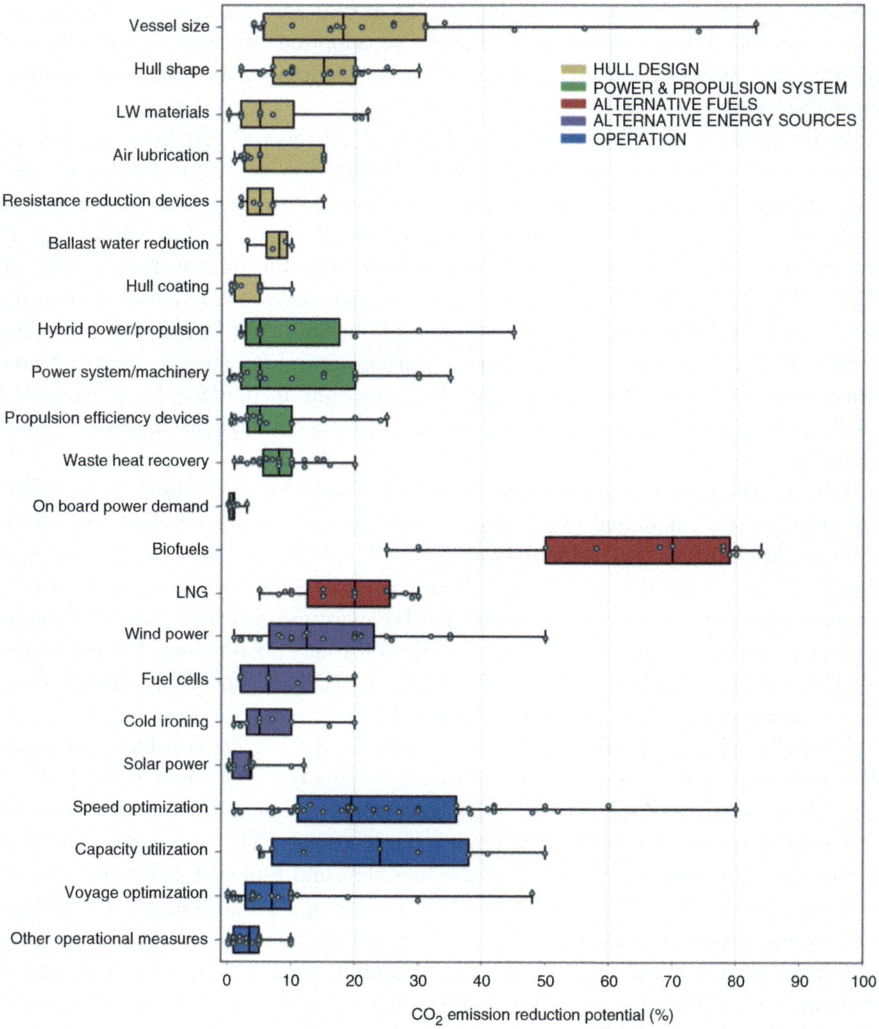

Fig. 10.3 CO_2 emission reduction potential from individual measures, classified in five main categories of measures. (Source: Bouman et al. (2017))

(LNG), liquified petroleum gas (LPG), hydrogen, methanol, ammonia, biofuels (ethanol, biodiesel, etc.), and fuel cells. In their view, until new technologies (batteries, synthetic fuels, synthetic biofuels, or others) that provide a "quantum leap" become economically viable, current conventional liquid fuels have the smallest GHG footprint.

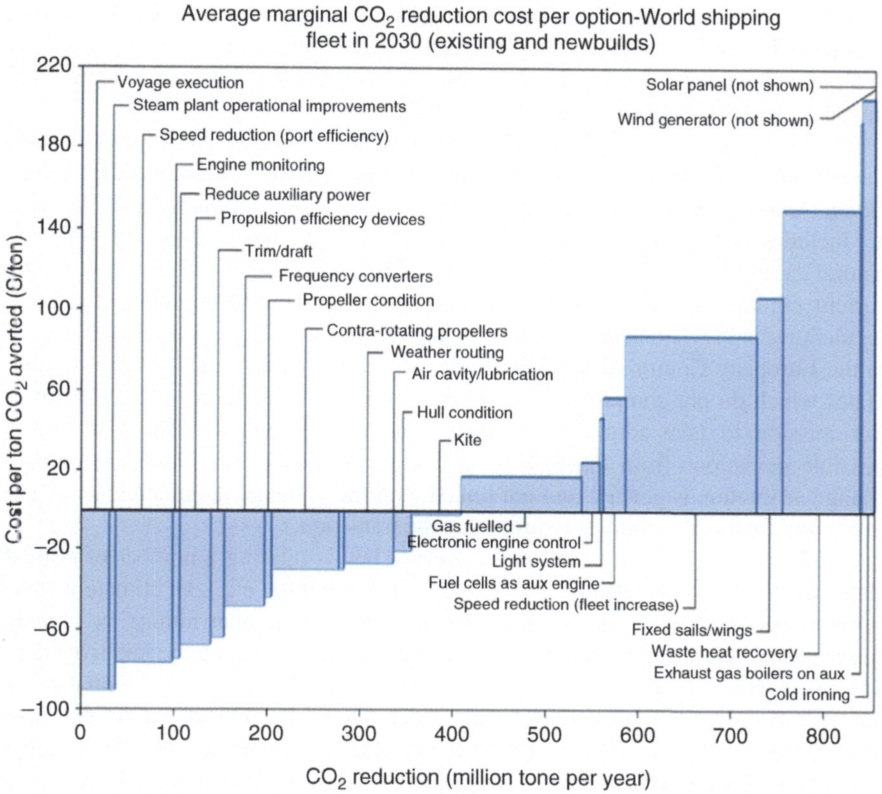

Fig. 10.4 Average marginal CO_2 reduction cost per option. (Figure adapted from the study by Eide et al. (2011). Source: Wan et al. (2018))

10.4.5 *European Green Deal and European Climate Law*

On December 2019, European Commission (EC) announced the introduction of the European Green Deal (EGD), which aims to boost the implementation of the Agenda 2030 and of the 17 SDGs. Europe aims to become the world's first climate-neutral continent by 2050 increasing the EU's greenhouse gas emission reductions target for 2030 to at least 50% and toward 55% compared with 1990 levels (European Commission 2019). The EGD aims at the reformation of four sectors of the economy, namely, energy, buildings, industry, and mobility. The majority (75%) of the EU's GHG emissions are associated with the energy sector, while transport represents 25% of EU's GHG.

In order for a just and inclusive transition to be achieved, the European Commission commits to support companies' transition to green and clean technologies. A 90% reduction in the transport emissions is targeted by 2050, while the supply of sustainable alternative transport fuels (e.g., biofuels and hydrogen) is boosted and promoted in shipping including other forms of transport, such as aviation and road transport. The European Community Shipowners' Association (ECSA) support European Commission EGD, and specifically it encourages the evaluation of the roll out of infrastructure for the delivery of alternative (non-fossil) fuels and the transition to no pollution in ports (ECSA 2020).

The European Green Deal was followed up with a proposal for European Climate Law on March 2020 aiming at writing into law the objective set out in the European Green Deal (e.g., Europe to become the first climate-neutral continent by 2050). This act includes cutting emissions, investing in green technologies, and protecting the natural environment (European Commission 2020a). Despite its positive nature, the current climate law neither sets an ambitious goal for 2030 nor does it refer to regulations and revisions needed for its achievement. Also, it gives significant power to the European Commission without allowing it to impose sanctions on Member States which do not comply with the respective recommendations of the European Commission to take additional measures and change policies that will correct possible deviations from the path to achieving the goals. The article on climate change adaptation is generic and not linked with the systemic documentation of the needs and financial resources required for the transition.

In addition, the EC proposal on the climate law is missing a number of critical elements. Besides core sectors (e.g., energy, transportation, etc.), all climate-related components need to be included in the decarbonization plan aiming at climate neutrality, namely, waters, underground waters, biodiversity, forestry, and livestock. The decarbonization plan should also include the time factor, i.e., individual goals for each objective (short-medium-long), which should be achieved in all environmental components. In other words, for the climate law to be applicable in the maritime sector it needs to identify key time-linked targets for ships, which ensure that the greater objective of climate neutrality will be reached.

Special care must be taken to create equal conditions for competition between EU companies and non-EU companies, mitigating the risk of "carbon leakage." More specifically, a clear reference is made to the obligation to formulate a relevant policy that ensures that the European Union's relations with third countries take into account their commitment and contribution to climate neutrality. This policy and tools could include, for example, linking aid and funding programs to complying with agreed climate change targets, or providing technology exchange and know-how only to compliant countries. Particular care is needed to support the shipping industry to cope with the challenge of its transformation aiming at the harmonization with the climate neutrality objective.

Climate law lacks an explicit description of the financial mechanisms that will be essential for climate neutrality achievement, which is the main objective of this law. In particular, the EU's financial gap for achieving the 2030 energy and climate targets is estimated at €180 billion a year, increasing the pressure on the climate law to include a financial plan of implementation. Therefore, many initiatives are needed to create the appropriate financial framework for the raising of the necessary funds while taking advantage of the roadmap for sustainable financing. In addition, the EU should consider imposing a carbon tax as well as strengthening the EU Emissions

Trading System (EU ETS), whose prices have fallen sharply due to the COVID-19 crisis (European Commission 2020b).

Since November 2014, rules related to the Effort Sharing Decision are being implemented by the Union, building up binding annual GHG emission targets for Member States for the period 2013–2020. The Effort Sharing Decision concerns emissions from non-included EU ETS sectors, such as transport, buildings, agriculture, and waste. In detail, the transport sector, as described in the Effort Sharing Decision, does not include aviation and international maritime shipping, which are large and growing sources of GHG emissions, due to their energy intensity and market share in global trade. This need is also emphasized by Directive (EU) 2018/410 of the European Parliament and of the Council, which highlights the need for EU ETS to act on shipping emissions as well as all other sectors of the economy.

10.4.6 Market-Based Mechanisms for GHG Mitigation

Several market-based mechanism proposals have been submitted to the Maritime Environment Protection Committee (MEPC). A sector-wide cap on net emissions from international shipping and a trading system alongside this was recommended by Norway. A similar proposal was suggested by France but also included an auction design. An Emissions Trading System was proposed by the UK with an initial phase including offsets for emissions. The US Ship Efficiency and Credit Trading preferred a mandatory energy efficiency standard enforced via an efficiency credit trading program. Importantly, in February 2017 the EU parliament voted to include shipping into the EU-ETS as of 2023 if there is an absence of action from the IMO by 2021. This causes concern among industry stakeholders that such a regional MBM would create distortions and may not lead to reduced CO_2 emissions, though the intent is to catalyze global action (Balcombe et al. 2019).

Broadly speaking market-based approaches can be divided into three categories: environmental price control approach, environmental quantity control approach, and subsidies. The environmental price approach can involve emissions charges or charges on fuels. The latter means that some opportunities for decoupling are lost, e.g., carbon capture, but may be easier to enforce. Kosmas and Acciaro (2017) consider bunker levy schemes for GHG emission reductions in the form of a unit-tax per ton of fuel and an ad valorem tax. While recognizing that MBMs do not seem to be up for discussion in the foreseeable future, Psaraftis (2019) sees the idea of a significant bunker levy at a global level worth pursuing. He points to how higher fuel prices in Europe and Japan have had a significant impact on the fuel efficiency of their cars relative to the USA. Importantly a levy (or any charge resulting from tax or permits) should not be confined to marine transport as this could lead to a modal shift to land-based modes that are generally greater emitters of GHG.

The emission quantity control approach includes credit programs that provide operators with credits to if they undertake or support activities that reduce emissions. Benchmarking trading programs sets an average emissions level that should not be exceeded and usually allow for offsetting as opposed to elimination of emissions. A cap-and-trade program sets a total aggregated cap on emissions and allocates emission allowances that can then be traded by emitters.

Subsidies can be used to provide direct financial support for mitigation. Under the Freight Technology Incentives Program, subsidies are provided by Transport Canada to encourage the employment of energy-efficient technologies (Nikolakaki 2013).

The global application of market-based measures is essential to avoid carbon leakage and competitive distortions especially given the relative ease with which ships are able to change their legal jurisdiction and register flags of convenience with more lenient carbon regulation. A maritime ETS or a carbon tax, or some hybrid system of emission trading with a price floor and/or ceiling, could provide cost-efficient emission reductions allowing for the fullest range of responses by ship-owners. An additional advantage of a tax or auction of permits is that the funds raised could be used to support technological innovation, cover administrative costs, and be used to redistribute funds toward developing countries and climate change funds. A key challenge for such a system is the costs of administering, monitoring, and enforcing these measures. Given the myriad of options available for mitigation in the shipping industry, market-based mechanisms have the advantage of not attempting to pick the technological or operational fix. On the other hand, a short-term option like LNG may require a combination of subsidies and port dues to effectively accelerate the large capital infrastructural costs involved.

10.5 Sustainability Initiatives in Maritime Transport

Beyond regulatory measures and IMO strategies, there have been a number of government-led initiatives for sustainability in transport more generally and maritime transport in particular that have emerged. The 2011 EC White Paper on Transport defines a strategy toward competitive and resource-efficient transport systems and a number of objectives and targets that relate to the logistics chains, promotion of more energy-efficient modes of transport at a larger scale, and reduction in emissions. Another example is the 2012 China Green Freight Initiative (CGFI) that seeks to improve fuel efficiency, reduce CO_2 emissions, and adopt cleaner technologies in freight transport. Other states have also promoted sustainability targets and measures (Benamara et al. 2019).

There are also numerous industry-led voluntary actions and initiatives. Maersk, for instance, has developed an "eco voyage" maritime software tool which can help cut fuel costs and make a voyage plan resulting in minimum fuel consumption. CMA CGM decided to equip its future giant containerships with engines using LNG meant to bring about large reductions in pollution emissions. Examples of voluntary self-regulation in maritime transport include the Clean Cargo Working Group that provides tools to help understand and manage sustainability impacts, the Sustainability Shipping Initiative that brings leading companies to promote a sustainable future, and Eco-Ships that involves investing and ordering a new generation of vessels that are eco-friendly and at the same time fuel efficient. For more examples of voluntary self-regulation in maritime transport, see Benamara et al. (2019) and Lun et al. (2015). Industry players with a role in promoting sustainability often in the form of enforcing international commitments to standards include entities as diverse as the International Chamber of Shipping (ICS), the International Association of dependent Tanker Owners (INTERTANKO), and the International Association of Ports and Harbors (IAPH) (Benamara et al. 2019).

With growing institutional pressures, and hopefully heightened awareness, shipping firms are likely to increasingly engage in sustainability management. This is evinced in part through numerous voluntary initiatives. Beyond government- and industry-led initiatives, shipping firms are responding to the rapidly evolving regulatory challenges as well as the institutional pressures from civil society and investors. The environmental, social, and governance (ESG) rating industry is putting pressure on international and domestic companies to improve their sustainability profile. ESG reports, ratings, and indices are increasingly relied upon by institutional investors, asset managers, financial institutions, and other stakeholders to assess and measure company sustainability performance. There are many ESG data providers like Bloomberg ESG Data Service, Sustainalytics Company ESG Reports, and Dow Jones Sustainability Index (DJSI), to name a few. Huber and Comstock (2017), Siew (2015), and Olmedo et al. (2010) offer an overview of the sustainability rating indices and agencies. In response to the greater demand of stakeholders for greater transparency in sustainability matters, shipping companies may undertake sustainability reporting on their own and in conjunction with third-party certification agencies. For an overview of corporate sustainability reporting tools, see Siew (2015). For a company to achieve good sustainability ratings or to gain certification for (dimensions of) sustainability, ultimately it needs to adjust or fundamentally alter its strategic vision and management approach. There is a vast and rapidly growing literature on sustainable corporate management (Epstein and Buhovac 2017; Modak 2018; Brockett and Rezaee 2012; Lambin and Thorlakson 2018; Chrun et al. 2016). There is less literature focusing squarely on sustainable company management in the context of the marine transport industry or the link between such sustainability practices, institutional pressures, and performance outcomes of companies. Lun et al. (2015) raise this issue in a book focusing on green shipping management. They point to many firms that are placing importance on environmental protection when performing shipping activities such as mega carriers (e.g., Hapag-Lloyd, APL, K Line, Maersk, NKY, and OOCL) and giant shippers (e.g., IKEA, Mattel, Nike, Home Depot, and HP) that are members of the Clean Cargo Working Group looking to integrate sustainability business principles into transport management. In the very broad context, Lambin and Thorlakson (2018) look at sustainability standards and how the overall interaction between private actors, civil society, and governments is reshaping global environmental governance.

10.6 Conclusions

Maritime transport has a critical role in addressing the sustainability challenges of our times. It plays a key role in international trade, providing market access and linking communities. "Safe, secure, energy-efficient, affordable, reliable, low-carbon, environmentally friendly, climate-resilient and rule-based maritime transport systems contribute to achieving an economically efficient, socially equitable and environmentally sound development" (Benamara et al. 2019). The new regulatory challenge posed by the sulfur cap in 2020 has generated substantial uncertainty in the shipping industry. While the shipping industry is focusing on the sulfur cap, the greatest challenge it has likely ever faced is the need to find the effective means of decarbonizing in line with global commitments. The speed of the required transition along with the relative difficulty of technological options vis-a-vis other sectors of the economy makes this a particularly demanding endeavor.

A number of government-led initiatives indicate a growing awareness of the shipping challenge, while initiatives at the level of industry and companies suggest a new reckoning of corporate responsibility. The International Maritime Organization will have a critical role to play in determining the right approach for decarbonization policy. Market-based mechanisms could potentially play an important role though they are still far from the center of the debate. They can incentivize the low carbon transition, spurring innovation across CO_2 emissions options and providing needed funding both for innovation and supporting developing economies address the heightened burdens of the transition. They are likely however to be one of many measures, regulations, and initiatives needed for the task. Scaling up financial resources and investments will also be an important enabler. This is a role that can be undertaken by regional and national development banks, e.g., the European Investment Bank (EIB) and ING signed an agreement to support the European shipping market with 300 million worth of green investment. Green bonds are another potential instrument for large infrastructural investments.

Enhancing the sustainability of the maritime transport will require a multisector approach involving governments, transport industry, financial institutions, academia, and civil society. The inherently international nature of maritime transport would seem to make it especially suited for global challenges, but it is also a potential weakness in that most governance institutions and their means of enforcing law and regulation are national in nature. Besides the ambitious goals of the European Green Deal and the European Climate Law of achieving climate neutrality by 2050 in alignment with the Agenda 2030 and IMO regulation, there is a lack of depth in the individual targets that need to be met by all parties that can conrtribute to the low carbon transition. Shipping industry, as well as other sectors, needs to be specifically mentioned and targeted in the short-medium-long term in these agendas; otherwise, the overarching goal of reducing GHG by 2050 will not be achieved. Furthermore, financial mechanisms that will be essential for climate neutrality achievement need to be explicitly stated, while maritime transport emissions should be included in the EU ETS if the IMO doesn't take effective action by 2021.

We have focused here primarily on some of the environmental dimensions of sustainable shipping, partly because of the particular historic junction we are at, but there are many other aspects of sustainable shipping that need to be part of our overall effort to achieve the Sustainable Development Goals. For instance, while more women have been entering the shipping industry in all roles, and efforts to advance their role have been made, the level of women's participation in the maritime industry remains low at an estimated 2%, and patterns of job segregation persist (UNCTAD 2018). Sustainability is a very broad and sometimes ambiguous concept, but it captures societal values and shapes our vision with a persistence similar to that of those other familiar concepts like democracy, justice, and liberty that have driven change throughout human history.

References

Alvik, S., Eide, M. S., Endresen, Ø., Hoffmann, P., & Longva, T. (2010). *Pathways to low carbon shipping. Abatement potential towards 2030.* DNV.

Asariotis, R., Benamara, H., & Mohos-Naray, V. (2017). *Port industry survey on climate change impacts and adaptation.* UNCTAD/SER.RP/2017/18.

Balcombe, P., Brierley, J., Lewis, C., Skatvedt, L., Speirs, J., Hawkes, A., et al. (2019). How to decarbonise international shipping: Options for fuels, technologies and policies. *Energy Conversion and Management, 182*, 72–88.

Benamara, H., Hoffmann, J., & Youssef, F. (2019). Maritime transport: The sustainability imperative. In H. Psaraftis (Ed.), *Sustainable shipping: A cross-disciplinary view*. Denmark: Springer.

Bouman, E. A., Lindstad, E., Rialland, A. I., & Strømman, A. H. (2017). State-of-the-art technologies, measures, and potential for reducing GHG emissions from shipping – A review. *Transportation Research Part D: Transport and Environment, 52*, 408–421.

Brockett, A., & Rezaee, Z. (2012). *Corporate sustainability*. Wiley.

Chrun, E., Dolšak, N., & Prakash, A. (2016). Corporate environmentalism: Motivations and mechanisms. *Annual Review of Environment and Resources, 41*(1), 341–362.

Directive (EU) 2018/410 of the European Parliament and of the Council of 14 March 2018 amending Directive 2003/87/EC to enhance cost-effective emission reductions and low-carbon investments, and Decision (EU) 2015/1814. https://eur-lex.europa.eu/legal-content/EN/TXT/?uri=CELEX:32018L0410

DVN GL. (2017). *Low carbon shipping towards 2050.* Norway. https://www.dnvgl.com/publications/low-carbon-shipping-towards-2050-93579

ECSA. (2020). *POSITION PAPER. A green deal for the European shipping industry.* https://www.ecsa.eu/sites/default/files/publications/2020%20ECSA%20Position%20Paper%20-%20A%20Green%20Deal%20for%20the%20European%20shipping%20industry.pdf

EIA. (2016). *International energy outlook 2016.* Washington, DC: U.S. Energy Information Administration.

EIA. (2017). *International energy outlook 2017.* Washington, DC: U.S. Energy Information Administration.

Eide, M. S., Longva, T., Hoffmann, P., Endresen, Ø., & Dalsøren, S. B. (2011). Future cost scenarios for reduction of ship CO_2 emissions. *Maritime Policy & Management, 38*(1), 11–37.

Endres, S., Maes, F., Hopkins, F., Houghton, K., Mårtensson, E. M., Oeffner, J., et al. (2018). A new perspective at the ship-air-sea-interface: The environmental impacts of exhaust gas scrubber discharge. *Frontiers in Marine Science, 5*.

EPA, U. S. (2013). *Clean air markets: Allowance trading*. Environmental Protection Agency. http://www.epa.gov/airmarkets/trading/

EPA, U. S. (2016). *National port strategy assessment: Reducing air pollution and greenhouse gases at U.S. Ports*. Environmental Protection Agency: Office of Transportation Air Quality. EPA-420-R-16-011. http://www.epa.gov/airmarkets/trading/

Epstein, M. J., & Buhovac, A. R. (2017). *Making sustainability work*. Routledge.

European Commission. (2019) *The European Green Deal, Brussels, 11.12.2019 COM (2019) 640 final Communication from the commission to the European Parliament, the European Council, the council, the European Economic and Social committee and the committee of the regions*. https://ec.europa.eu/info/sites/info/files/european-green-deal-communication_en.pdf

European Commission. (2020a). *European climate law – achieving climate neutrality by 2050*. https://ec.europa.eu/info/law/better-regulation/have-your-say/initiatives/12108-Climate-Law

European Commission. (2020b). *EU Emissions Trading System (EU ETS)*. https://ec.europa.eu/clima/policies/ets_en

Eyring, V., Kohler, H. W., Van Aardenne, J., & Lauer, A. (2005). Emissions from international shipping: 1. The last 50 years. *Journal of Geophysical Research: Atmospheres, 110*, D17.

Halff, A., Younes, L., & Boersma, T. (2019). The likely implications of the new IMO standards on the shipping industry. *Energy Policy, 126*, 277–286.

Huber, B. M., & Comstock, M. (2017). *ESG reports and ratings: What they are, why they matter*. Harvard Law School Forum on Corporate Governance and Financial Regulation. https://www.reprisk.com/content/6-news/1-media-coverage/385-esg-reports-and-ratings-what-they-are-why-they-matter/esg-reports-and-ratings-what-they-are-why-they-matter.pdf

ICCT. (2011). *Reducing greenhouse gas emissions from ships cost effectiveness of available options*. International Council for Clean Transportation. https://www.theicct.org/sites/default/files/publications/ICCT_GHGfromships_jun2011.pdf

ICS. (2016). *Proposal to develop an 'intended IMO determined contribution' on CO 2 reduction for international shipping*. International Chamber of Shipping. MEPC 69, 7(1). http://www.ics-shipping.org/docs/default-source/Submissions/IMO/proposal-to-develop-an-intended-imo-determined-contribution-on-co2-reduction-for-international-shipping.pdf?sfvrsn=0

IMO. (2014). *Third IMO greenhouse study 2014. Executive summary and final report*. London: International Maritime Organization. http://www.imo.org/en/OurWork/Environment/PollutionPrevention/AirPollution/Documents/Third%20Greenhouse%20Gas%20Study/GHG3%20Executive%20Summary%20and%20Report.pdf.

IMO. (2017). *Considerations of how to progress the matter of reduction of GHG emissions from ships. Note by the secretariat*. London: International Maritime Organization. ISWG-GHG1/2.

IMO. (2018). *Report of the Marine Environment Protection Committee on it its seventieth session, road map for developing a comprehensive IMO strategy on, reduction of GHG emissions from ships*. International Maritime Organization. MEPC 70/18/Add.1, Annex 11. https://www.mardep.gov.hk/en/msnote/pdf/msin1707anx3.pdf

International Gas Union. (2017). *Enabling clean marine transport*. http://www.igu.org/sites/default/files/nodedocument-field_file/IGU_A4_CleanMarineTransport_Final%20March%202017_3.pdf

Kosmas, V., & Acciaro, M. (2017). Bunker levy schemes for greenhouse gas (GHG) emission reduction in international shipping. *Transportation Research Part D: Transport and Environment, 57*, 195–206.

Lambin, E. F., & Thorlakson, T. (2018). Sustainability standards: Interactions between private actors, civil society, and governments. *Annual Review of Environment and Resources, 43*, 369–393.

Lun, Y. H. V., Lai, K.-h., Wong, C. W. Y., & Cheng, T. C. E. (2015). *Green shipping management*. Springer.

Maloni, M., Paul, J. A., & Gligor, D. M. (2013). Slow steaming impacts on ocean carriers and shippers. *Maritime Economics & Logistics, 15*(2), 151–171.

McKinnon, A. (2016). *UNCTAD workshop on sustainable freight transport and finance*. Nairobi.

Mikelis, N. (2019). Ship recycling. In H. Psaraftis (Ed.), *Sustainable shipping: A cross-disciplinary view* (pp. 203–248). Denmark: Springer.
Modak, P. (2018). *Environmental management towards sustainability*. CRC Press.
Nikolakaki, G. (2013). Economic incentives for maritime shipping relating to climate protection. *WMU Journal of Maritime Affairs, 12*(1), 17–39.
OECD. (2017a). *CO2 emissions from fuel combustion: Overview*. Paris: Organisation for Economic Co-operation and Development/International Energy Agency.
OECD. (2017b). *ITF transport outlook*. Paris: Organization for Economic Co-operation and Development.
Olmedo, E. E., Torres, M. J. M., & Izquierdo, M. A. F. (2010). Socially responsible investing: Sustainability indices, ESG rating and information provider agencies. *International Journal of Sustainable Economy, 2*(4), 442.
Psaraftis, H. (Ed.). (2019). *Sustainable shipping: A cross-disciplinary view*. Denmark: Springer.
Psaraftis, H. N., & Zachariadis, P. (2019). The way ahead. In H. Psaraftis (Ed.), *Sustainable shipping: A cross-disciplinary view* (pp. 203–248). Denmark: Springer.
Siew, R. Y. (2015). A review of corporate sustainability reporting tools (SRTs). *Journal of Environmental Management, 164*, 180–195.
Traut, M., Larkin, A., Anderson, K., McGlade, C., Sharmina, M., & Smith, T. (2018). CO2 abatement goals for international shipping. *Climate Policy, 18*(8), 1066–1075.
UNCTAD. (2015). *Sustainable freight transport systems: Opportunities for developing countries. Note by the UNCTAD secretariat.* New York: United Nations Conference on Trade and Development. TD/B/C.I/MEM.7/11. https://unctad.org/meetings/en/SessionalDocuments/cimem7d11_en.pdf
UNCTAD. (2018). *Review of maritime transport 2018*. New York: United Nations Conference on Trade and Development. https://unctad.org/en/PublicationsLibrary/rmt2018_en.pdf.
UNFCCC. (1997). *Kyoto protocol to the United Nations framework convention on climate change adopted at COP3 in Kyoto, Japan, on 11 December.* https://unfccc.int/resource/docs/cop3/07a01.pdf
Wan, Z., el Makhloufi, A., Chen, Y., & Tang, J. (2018). Decarbonizing the international shipping industry: Solutions and policy recommendations. *Marine Pollution Bulletin, 126*, 428–435.

Chapter 11
New Challenges and Opportunities for Sustainable Ports: The Deep Demonstration in Maritime Hubs Project

Vera Alexandropoulou, Phoebe Koundouri, Lydia Papadaki, and Klimanthia Kontaxaki

Abstract Environmental challenges related to ports are twofold, namely, the effects of maritime transport on the environment (e.g. pollution, CO_2 emissions) and conversely the environmental impact on maritime transport, e.g. climatic variability and change. This chapter (This work has received funding from the European Union's European Institute of Innovation and Technology under grant agreement N° 201,166.) presents an overview of main challenges faced today, to engage port proactively take the responsibility of providing reward schemes or green certificates to complied ships and to identify key indicators in measuring GHG emissions. European Union has put into force a number of Directives and Regulations aiming to incentivise port and shipping companies to commit to comply with environmental standards. The IMO 2020 regulation, bringing the sulphur cap in fuel oil for ships down from 3.50% to 0.50%, is expected to bring significant benefits for human health and the environment, while the European Green Deal, the most ambitious action plan of European Union, aims at increasing the EU's greenhouse gas emission

V. Alexandropoulou (✉)
Thalassa Foundation, Athens, Greece

Vera Alexandropoulou Law, Athens, Greece
e-mail: vera.alexandropoulou@thalassafoundation.com

P. Koundouri
School of Economics and ReSEES Laboratory, Athens University of Economics and Business, Athens, Greece

United Nations Sustainable Development Solutions Network Europe, Paris, France

EIT Climate–KIC Hub Greece, ATHENA Research and Innovation Center, Athens, Greece

ICRE8: International Centre for Research on the Environment and the Economy, Athens, Greece

L. Papadaki
United Nations Sustainable Development Solutions Network Greece, Athens, Greece

EIT Climate-KIC Hub Greece, ATHENA Research and Innovation Center, Athens, Greece

K. Kontaxaki
Vera Alexandropoulou Law, Athens, Greece

P. Koundouri (ed.), *The Ocean of Tomorrow*, Environment & Policy 57,
https://doi.org/10.1007/978-3-030-56847-4_11

reductions target for 2030 to at least 50% compared with 1990 levels, creating the most ambitious package of measures, accompanied by an initial roadmap of key policies in cutting-edge research and innovation, in green technologies and sustainable solutions. Among them, Deep Demonstrations by EIT Climate-KIC using systems innovation approach aim at the decarbonisation of the European ports and the sustainable transformation of their key elements.

Keywords Sustainable ports · European Green Deal · Maritime transport · Ports regulation · Deep Demonstration · Environmental policy

11.1 Introduction

Sea ports are major hubs of economic activity which may also involve severe environmental pollution in coastal urban areas. The port industry along with the shipping industry constitutes a key node in the international supply chain taking into consideration that over 80% of volume (70% of value) of world's merchandise trade is carried by sea (port to port). Due to increasing global trade, transport of goods through ports has been steadily increasing and will likely continue to increase in the future. The United Nations Conference on Trade and Development is projecting an annual average growth rate of 3.4% for the Maritime Trade period 2019–2024 (UNCTAD 2019).

Ports and shipping are intrinsically linked – as such, efforts to reduce maritime emissions need to extend beyond seagoing ships alone. IMO's MARPOL Annex VI (2010) regulations on air pollution and energy efficiency are aimed at ships; however, it is clear that in order for port emissions to be reduced, emissions from all port-related emission sources need to be addressed.

Environmental challenges relating to ports are twofold, namely, the effects of maritime transport on the environment (e.g. pollution, CO_2 emissions) and conversely the environmental impact on maritime transport (e.g. climatic variability and change, (CV&C)) (Asariotis et al. 2018). In this regard, it is important to address the global challenges effectively, in the light of the Paris Agreement and the 2030 UN Sustainable Development Agenda. Reducing the sources of GH emissions and of marine pollution emanating from the port industry is of growing importance and source of anxiety for port authorities, policy makers, port users and the local communities (Acciaro et al. 2014).

11.2 Ports Needs Towards Environmental Sustainability

This analysis focuses only on the environmental dimension of port sustainability although the port sustainability literature engages in the social and the economic aspect as well (Özispa and Arabelen, 2018). As environmental concerns over managing seaports are gaining utmost importance, evaluating the greenness of the port draws a serious academic and research attention. A set of indicators, both qualitative as well as quantitative, visualising the environmental sustainability perspective has been identified under several scientific studies (Puig et al. 2014). Indicators which formulate the port's environmental sustainability constitute inter alia, waste management and handling, ballast water and water conservation and quality, air quality and reduction of emissions, noise control, energy efficiency and transition to cleaner energy.

As the shipping industry and international trade increase, ports are improving maritime infrastructure and enhancing port facilities with smarter, more intelligent designs with the help of technology. By going digital, connectivity and automation may reduce environmental footprints of the port industry along with intelligent transport systems, which have a significant potential to reduce CO_2 emissions. Environmental reporting is also becoming increasingly important for ports in the face of growing environmental concerns and stakeholder pressure from market players, public bodies and social interest groups.

Evaluating air pollution impacts of ports requires consideration of numerous sources, including marine vessels, trucks, locomotives and off-road equipment used for moving cargo. The air quality impacts of ports are significant, with particularly large emissions of diesel exhaust, particulate matter and nitrogen oxides. Approaches to mitigation encompass a range of possibilities from currently available, low-cost approaches to more significant investments for cleaner air, such as restrictions on truck idling and the use of low-sulphur diesel fuel; the latter includes shoreside power for docked ships and alternative fuels (Bailey and Solomon 2004).

The IMO (2020) regulation, bringing the sulphur cap in fuel oil for ships down from 3.50% to 0.50%, is expected to bring significant benefits for human health and the environment. Enforcement, compliance with and monitoring of the new sulphur limit is the responsibility of state parties, both in their capacity as flag states or port states, to the International Convention for the Prevention of Pollution from Ships (MARPOL 1973), as modified by the Protocol of 1978 (MARPOL 73/78), Annex VI. Ships found to be in non compliance can be detained by port state control and/or sanctions may be imposed for violations. Furthermore, the additional amendment to MARPOL 73/78 which enters into force on 1 March 2020 prohibits not only the use but also the carriage of noncompliant fuel oil for combustion purposes for propulsion or operation on board a ship, unless it is fitted with a scrubber, which is an exhaust gas cleaning system.

A variety of further measures are suggested towards the reduction of port emissions such as introducing differentiated port dues, providing onshore power supply/'cold ironing', switching to low-sulphur fuels at berth and establishing speed limits in ports. In addition, the improvement of the exchange of information between ports and ships so that ships are able to sail at optimal speed (virtual arrival) is of great importance. Another potential measure is giving preferential treatment to harbour crafts with engines that meet stringent emissions standards while, on the other hand, strengthening port state control inspection regimes for visiting ships, relating to compliance with MARPOL, Annex VI. Finally, the designation of additional emission control areas, leading to stricter environmental emission standards enforced at certain ports (ships going through them should use fuel with sulphur content lower than 0.10% (below the 0.5% limit applicable on 1 January 2020)), could make a significant difference (UNCTAD 2019).

11.3 Systems Innovations Approach to Engage Stakeholders in Co-designing and Implementing

Accordingly, ports are increasingly expected to align their performance with sustainability expectations, namely, to deliver optimum economic and social gains while causing minimum environmental damage through the ecosystem services approach (Boerema et al. 2017). As international trade and cargo volumes continue to grow, ports around the globe are looking to new technologies to help manage resources in a more sustainable and cost-effective manner through digitalization.

In view of the differences among ports and the changing nature of the environmental challenges that ports face, the establishment of an environmental management system is considered of utmost importance. A systematic approach to environmental management system enables the continuous identification of an individual port's priorities while it introduces a functional organisational structure that sets respective targets, implements measures, monitors impact and evaluates, reviews and takes corrective actions when and where necessary.To this effect, a methodology, the Self Diagnosis Method, has been designed to assess the performance of the environmental management in sea ports (Dabra et al. 2004). In this way ports can achieve and demonstrate continuous environmental improvement towards sustainability.

When it comes to the systems innovation, envisioning the desired future and learning from that become necessary. Visioning and backcasting are two pillars of the approach and should be done under a participatory approach. Since stakeholders have radically different world views and different frames for understanding the problem, you should incorporate their perspectives, even if they are wildly different to your own. Because of the different stakeholders' perspectives, they all have their own priorities and agendas. Involving them in the backcasting process will allow you to draw more than one plan from the same process. In complex and wicked problems, as sustainability is, the problem definition might come to focus after adopting a future vision. In such cases, the vision is the seed for the challenge and not a consequence of it. Visioning should be a participatory tool in which a large diversity of stakeholders ensures a richer and broader vision.

The adoption of new technologies in conjunction to the system innovation approach could prevent reaching the point of an environmental crisis in the port. This could be achieved through systematic and formal training of special scientists regarding the alternatives towards environmental sustainability, working closely with the practitioners to familiarise themselves with new technologies of controlling emissions considering localities and idiosyncrasies, explore onshore power supply (e.g. cold ironing) and consider it as an asset to be managed, better design pricing policies based on meeting international environmental standards and exploit the opportunities of renewable energy and energy communities as a primer source of port's energy.

Therefore, inviting all the interested parties to co-design and implement a commonly accepted solution adds value to the initiative as all of them work towards achieving the same goal. Such being the case, a mix of social and economic policy solutions is needed. All in all, the ecosystem services approach helps in the direction of supporting the agreed solution.

11.4 Deep Demonstrations for Net-Zero Emissions

Ports are places where multiple systems collide (shipping, energy, waste, tourism and other transport). Ports can either be emissions hotspots or hubs able to drive enormous change. In a phased way, EIT Climate-KIC[1] works with a small cohort of high-ambition port authorities in Valencia (Spain) and Piraeus (Greece) and Cyprus Ministry of Shipping to demonstrate how ambitious maritime hubs can be catalysts for reversing the fast-growing emissions from international shipping and trade using systems innovation approach.

Deep Demonstrations funded by EIT Climate-KIC start with a demand-led approach, working with organisations willing to take on the responsibility of acting as 'problem owners' – in Greece Piraeus Port Authority – committed to net-zero emissions and resilient futures (EIT Climate-KIC 2020). Deep demonstrations (Fig. 11.1) progress in tightly designed, iterative phases of rolling out systems innovation-as-a-service, aiming at the identification of the key actors to be involved, current status, vision, innovation needs, sustainable financial planning and ultimately at the alignment of all actors able to drive systems transition to a low-carbon emissions future. Deep Demonstrations are a circular approach in innovation implementation with a final goal: the holistic change of the port to sustainability.

Deep Demonstrations methodology is composed of four phases (Intent, Frame, Portfolio and Intelligence). Intent phase aims at analysing the current status of the port and identifying key stakeholders creating a consortium of key players able to drive the highly needed change and co-create a vision. This phase intends to develop a frame of reference for approaching innovation deliberately and systemically through a portfolio approach and sense-making in order to manage uncertainty and

[1]EIT Climate-KIC is a European knowledge and innovation community, working towards a prosperous, inclusive, climate-resilient society founded on a circular, zero-carbon economy. The EIT Climate-KIC is part of the European Institute of Innovation and Technology (EIT) and the EIT Community. The EIT is a body of the European Union and a global innovation leader, delivering world class solutions to societal problems. In particular, we aim to catalyse the rapid innovation needed across sectors by convening the brightest minds to tackle challenges, empowering leaders through capacity building and seed funding the most promising climate-positive businesses.

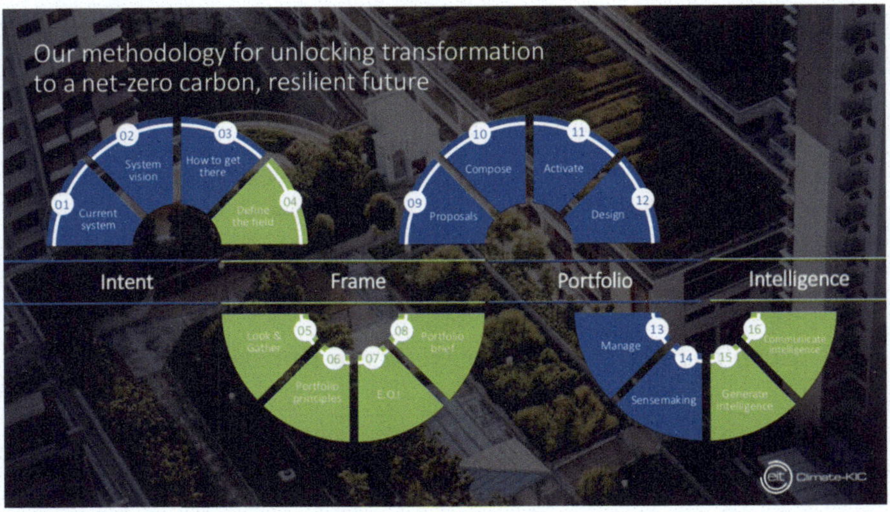

Fig. 11.1 Deep Demonstrations methodology. (Source: EIT Climate-KIC 2020)

generate options and intelligence from innovation experience. The Frame phase identifies and addresses ports' needs, cause and effect relationships and opportunities aiming at inviting innovation and research to meet these needs. The focus of the next phase, the Portfolio phase, is to raise awareness on the major challenges of the port and encourage diversity to ensure a spread of learning and connectivity and to enable the identification of multiplier effects and integrated solutions. The Intelligence phase is the ultimate objective of the Deep Demonstrations process. Intelligence is the outcome of sense-making and analytics drawing on innovation experience and learning from multiple different experiments deploying diverse leverage points.

The combination of all these phases in a circular manner can support challenge owners change mindset and action plan through understanding the interdependences among different actors and the common vision they can develop, which is beneficial delivering optimum economic, social and environmental gains.

11.5 Ports Role in Reducing the Global Carbon Footprint

Environmental sustainability in the port sector mainly relates to environmental performance and management. Environmental considerations may be different for each port, depending on the specific location and the characteristics of the port area. Seaport environmental management progressed over the last decades from a 'point-focused' seafront-based exercise to an integrated seaport area management concept. There is potential for further integration as seaports proactively act as facilitators of procedures and of communication between the different parties involved in the logistic chain. The concept of ports as facilitators refers to the contribution that ports can make in assisting the whole port community (including partners in the logistic chain) to deliver compliance with legislation, prevention of pollution, reduction and mitigation of environmental impacts, sustainable development and evidence of satisfactory performance. This resulted in the development of the port practice to include the sustainability performance, as part of the annual corporate social responsibility and financial report.

Greenhouse gas emissions from shipping currently represent around 2.6% of total global emissions, and without reduction measures, this share could more than triple by 2050. The International Maritime Organization (IMO) has set a target of reducing shipping CO_2 emissions from the shipping sector by 'at least' 50% by 2050 compared to 2008 levels. To achieve this, stringent measures now need to be put into place.

According to the analysis report published by the International Transport Forum (ITF) in 2018, ports play a significant role in reducing the global carbon footprint of maritime shipping, and consequently portside measures can significantly add to the environmental performance of shipping and the decarbonisation of maritime transport (OECD/ITF 2018). Currently 28 of the 100 world's largest ports (in terms of total cargo volume handled) offer incentives for environmentally friendly ships: Some US ports offer fee reductions for ships reducing speed when approaching the port. The Panama Canal Authority provides priority slot allocation to greener ships. Spain includes environmental incentives in the tender and licence criteria for the towage services provided in ports. Shanghai has an emission-trading scheme that includes ports and domestic shipping. However, green incentives typically apply to the 5% of the ships calling at a port with an incentive scheme. Only five ports use CO_2 emissions as a substantial criterion for incentives. The report expands on port-based incentives for low-emission ships. It links port-based incentives to actual greenhouse gas emissions, moving to a more harmonised application of green port fees.

Notwithstanding the dearth of data, it is clear that the impact of port-based incentives on global shipping emissions is marginal. Currently only few ports use indices in which GHG emissions provide a substantial part of the index criteria. Yet, ports clearly play a hugely important role in helping the shipping sector to manage the transition to clean shipping. Port-based incentives for GHG emission mitigation could provide an important supporting role. The first lesson learned is therefore that ports are stakeholders in this context and that they are taking actions – to both incentivise cleaner ships and to increase the efficiency of their operations, which can also have an effect on shipping emissions. Furthermore, the existing port-based measures establish that market interventions are needed to reward clean performance. The fact that financial incentives have been chosen implies that there is support for flexible measures to drive behavioural change. However, more emphasis is needed on monitoring, reporting and verification of the impacts of these measures. More could also be done to enshrine the 'polluter pays' principle. Higher rates of differentiation between vessels based on their environmental performance could drive more and faster change. It is possible within the policies to differentiate fees according to type of vessel enabling the economic activities that can afford to pay to take more of the responsibility for acting.

A project on the Environmental Impacts of International Shipping and the role of ports, that took place under the aegis of OECD (OECD 2011), showed that while it is difficult to identify 'best practices' for all the environmental impacts that port activities generate, the introduction of shoreside electricity supply ('cold ironing') is identified as a specific measure that would have the advantage of reducing several negative impacts simultaneously, such as SO2, NOx and particulates emissions, noise and, possibly, CO_2 emissions. In states where electricity generation is covered by a 'cap-and-trade' system for CO_2 emissions (e.g. in some EU states), the latter would be the case, regardless of how the electricity used to supply the ships is produced, as long as the 'cap' of the trading system remains unchanged. However, an important obstacle to a broader use of shoreside electricity is that electricity systems vary between states, both in terms of voltage, frequency and pricing. And it is not enough to make shoreside electricity available, unless ships are obliged to use it; they have few incentives to do so.

11.5.1 Environmental Ship Index (ESI)

On 12 May 2017 the International Association of Ports and Harbors decided to set up a World Ports Sustainability Program (WPSP) guided by the 17 UN SDGs in order to enhance and coordinate future sustainability efforts of ports worldwide and foster international cooperation with partners in the supply chain. The WPSP aims to demonstrate global leadership of ports in contributing to the Sustainable Development Goals of the United Nations. The cooperation between port professionals, academic researchers and specialist organisations has proven to be a potent mix in terms of delivering a functional framework of cost-effective solutions developed to implement policies and produce continuous improvement of the port environment.

The WPSP builds on the World Ports Climate Initiative that IAPH started in 2008 and extends it to other areas of sustainable development. Building on the output of the World Ports Climate Initiative, port community actors can collaborate in refining and developing tools to facilitate reduction of CO_2 emissions from shipping, port and landside operations. In addition, they can take initiatives to enable energy transition, improve air quality and stimulate circular economy. The American Association of Port Authorities (AAPA), the European Sea Ports Organisation (ESPO), the International Association of Cities and Ports (AIVP) and the World Association for Waterborne Transport Infrastructure (PIANC) signed up as strategic partners of the World Ports Sustainability Program.

One of the projects within WPSP is the Environmental Ship Index (ESI). The Environmental Ship Index (ESI) identifies seagoing ships that perform better in reducing air emissions than required by the current emission standards of the International Maritime Organization. The ESI evaluates the amount of nitrogen oxide (NOx) and sulphur oxide (SOx) that is emitted by a ship; it includes a reporting scheme on the greenhouse gas emission of the ship. The ESI is a perfect indicator of the environmental performance of ocean-going vessels and will assist in identifying cleaner ships in a general way. The index is intended to be used by ports to reward ships when they participate in the ESI in order to promote clean ships but can also be used by shippers and ship owners as their own promotional instrument. It should be noted that while the ESI database will provide a total score, the rewards can either be based on that total or on each of its constituent parts separately; for that purpose, those parts are appearing in the ship details. ESI is completely voluntary, and WPSP hopes that the global port community will assume its role in improving the maritime and port environment taking into account the priorities of ports with regard to the environmental performance of ships that ports wish to promote.

11.5.2 Port Emissions Toolkits

Toolkits to tackle ship and port emissions have been developed under the GEF-UNDP-IMO Global Maritime Energy Efficiency Partnerships (GloMEEP) project in collaboration with its partners, the Institute of Marine Engineering, Science and Technology (IMarEST) and the International Association of Ports and Harbors (IAPH).

The Port Emissions Toolkit includes two guides addressing the impact of air emissions from ports on the local and global environment which are as follows:

Guide No.1: Assessment of port emissions (GEF-UNDP-IMO GloMEEP Project and IAPH 2018a). The guide serves as a resource guide for ports intending to develop or improve their air pollutant and/or GHG emissions assessments by incorporating the latest emission inventory methods and approaches. It recognises that ships do not operate independently from shore-based entities in the maritime transportation system and that port emission considerations must extend beyond the ships themselves to include all port-related emission sources including seagoing vessels, domestic vessels, cargo handling equipment, heavy-duty vehicles, locomotives and electrical grid.

Guide No.2: 'Development of port emissions reduction strategies' (GEF-UNDP-IMO GloMEEP Project and IAPH 2018b) serves as a resource guide for ports intending to develop an emissions reduction strategy (ERS) for port-related emission sources. It describes the approaches and methods that can be used by ports to develop, evaluate, implement and track voluntary emission control measures that go beyond regulatory requirements.

By utilising these guides, national strategies can be developed which address emissions from the maritime sector as a whole – protecting public health and the environment while contributing to the fight against climate change.

11.6 EU Policies on Sustainable Ports

The EU has already in place an extensive and comprehensive regulatory environmental framework with which the European Ports' environmental policies must be aligned, indicatively: the 'Birds Directive' (Directive of the European Parliament and of the Council 2009/147/EC (2009) on the conservation of wild birds)), the Natura 2000 ecological network including all Special Protection Areas ('Habitats' Council Directive 92/43/EEC (1992)), Directive (EU) of the European Parliament and of the Council 2016/802/EU (2016) relating to a reduction in the sulphur content of certain liquid fuels, Regulation (EU) of the European Parliament and of the Council No 2018/1999 (2018) on the Governance of the Energy Union and Climate Action, Directive (EU) of the European Parliament and of the Council 2019/883/EU (2019) on port reception facilities for the delivery of waste from ships, Directive (EU) of the European Parliament and of the Council 2014/94/EU (2014) on Alternative Fuels Infrastructure (the AFID), Council Directive 2003/96/EC (2003) on the taxation of energy products and electricity, Regulation (EC) of the European Parliament and of the Council No 1013/2006 (2006) on shipments of waste, Directive of the European Parliament and of the Council 2008/98/EC (2008) on Waste (Waste Frameworh Directive), Directive of the European Parliament and of the Council 2008/50/EC (2008) on ambient air quality and cleaner air for Europe and Regulation (EC) of the European Parliament and of the Council No 1221/2009 (2009) on the voluntary participation by organisations in a community eco-management and audit scheme (EMAS). The Trans-European Transport Network (TEN-T) policy is based on Regulation (EU) of the European Parliament and of the Council No 1315/2013 (2013) aiming at sustainability through development of all transport modes in a manner consistent with ensuring transport that is sustainable and economically efficient in the long term; its contribution to the objectives of low greenhouse gas emissions, low-carbon and clean transport, fuel security, reduction of external costs and environmental protection and promotion of low-carbon transport with the aim of achieving by 2050 a significant reduction in CO_2 emissions, in line with the relevant Union CO_2 reduction targets.

Furthermore, new stricter environmental protection measures are on the way with the introduction of the European Green Deal (European Commission 2019). On 11 December 2019, the European Green Deal was communicated by the EU Commission, boosting a new strategy on implementing the United Nation's 2030 Agenda and the sustainable development goals, thereby increasing the EU's greenhouse gas emission reductions target for 2030 to at least 50% and towards 55% compared with 1990 levels. Becoming the world's first climate-neutral continent by 2050 is the most ambitious package of measures, accompanied with an initial roadmap of key policies ranging from ambitiously cutting emissions to investing in cutting-edge research and innovation, in green technologies and sustainable solutions. The Green Deal seeks a 90% reduction in the transport emissions by 2050, while it boosts the supply of sustainable alternative transport fuels – biofuels and hydrogen – which will be promoted in aviation, shipping and road transport. In addition, the European Green Deal purports to extend emissions trading to the maritime sector as well.

The overarching objective of the European Green Deal, aiming to reach net-zero greenhouse gas emissions by 2050, signifies an update of the EU's climate ambition for 2030, with a 50–55% cut in greenhouse gas emissions to replace the current 40% objective. To deliver these additional greenhouse gas emissions reductions, all relevant climate-related policies will be reviewed and potentially revised. To address these interlinked challenges, a zero-pollution action plan for air, water and soil will also be adopted. The 55% figure will be subject to a cost-benefit analysis of every EU law and regulation in order to align them with the new climate goals. Further decarbonising the energy system is critical to reach climate objectives in 2030 and

2050. The production and use of energy across economic sectors account for more than 75% of the EU's greenhouse gas emissions. In this regard, energy efficiency must be prioritised. A power sector must be developed that is based largely on renewable sources, complemented by the rapid phasing out of coal and decarbonising gas. At the same time, the EU's energy supply needs to be secure and affordable for consumers and businesses. Hence, the Renewable Energy Directive and the Energy Efficiency Directive as well as the Emissions Trading Directive will be revised accordingly. Most importantly the circular economy, including new waste and recycling laws, is erected as 'utmost priority' of the European Green Deal in the EU's effort to achieve net-zero carbon emissions by 2050.

Greening 'the port' means more than greening the transport side. All industry players in the port should have their agendas, goals and plans, and the port managing body must support the industries in the port in their pathways to a more sustainable future. This requires support for large investments in the provision of clean energy, connectivity of energy infrastructure networks and green grids, as well as support for innovative technological projects in and between ports. In addition, ports can also attract new investments in clean energy and technology and become centres of excellence and innovation, instead of being just energy 'takers'. Thanks to the presence of industry and the proximity to large urban agglomeration they also constitute an ideal location to develop circular economy projects.

11.7 The European Sea Ports Organisation (ESPO)

11.7.1 The Green Guide

Published back in 2012 by ESPO, the 'Green Guide' (ESPO 2012) towards excellence in port environmental management and sustainability defines the sector's vision on environmental governance and establishes a structured approach that European ports subscribe to while tackling their environmental liabilities. European ports commit to work towards continuously improving their environmental performance through focused action under five principles: exemplify, enable, encourage, engage and enforce. Overall, the ESPO Green Guide favours a bottom-up approach, in which port authorities are proactively taking responsibility and living up to the expectations of the community. It encourages ports to be responsible for their own initiatives, to benchmark their performance and to deliver science-based evidence of achievements. The ESPO Green Guide specifically addresses five major environmental issues, namely, air quality, energy and climate, noise, waste management and water quality. The guide is the outcome of the common work of port environmental professionals around Europe under the umbrella of the sustainability committee of ESPO and is accompanied by a best practice database that promotes existing port projects. It defines a common vision of the port sector on environmental sustainability, promotes the efforts of European port authorities in the field of environmental management and demonstrates evidence of the progress achieved by the sector over time. The guide provides guidance to ports in establishing and developing further their environmental management programmes, highlights the main environmental challenges that ports face and demonstrates response options, developing a common approach towards responsible action while respecting the diversity of ports, their competences and their abilities.

The European Sea Ports Organisation (ESPO) welcomed Europe's objective set out in the European Green Deal to become the world's first net-zero emission area by 2050 and to reduce emissions by 50% towards 55% compared with 1990 levels by 2030. On 19 February 2020, ESPO published its Position Paper on the European Green Deal objectives in ports (ESPO 2020). ESPO recognises the importance of LNG as a transition fuel and considers onshore power supply (OPS) as an important pillar of the future energy landscape. Consequently, investments in those technologies should be further encouraged. LNG's role as a transition fuel should be recognised. LNG has been one of the compliant fuels for shipping to meet the 0.1% sulphur cap in SECA areas (since 2015) and the overall 0.5% sulphur cap (effective as of 1 January 2020). Moreover, current LNG infrastructure may also be used for bio-LNG.

However, ESPO considers that there is still uncertainty, as to which clean fuels will be most suitable for each segment of shipping. ESPO therefore argues that any new legislation should retain the current flexibility for any clean fuels or technologies which provide equivalent solutions. New legislation should allow the uptake of a variety of clean fuels, rather than prescribing specific fuels for shipping.

A technology neutral approach is an absolute prerequisite to support innovation in different promising technologies. For ESPO, a goal-based approach with emission reduction standards accompanied by port roadmaps is the best way to ensure that Europe's greening objectives are achieved. An EU standard for the reduction of emissions at berths should initially address berths close to urban areas and should target specific segments of shipping such as cruise ships and ferries. The standard should be subsequently expanded to all segments of shipping taking into account progress on the development of clean technologies. Any technologies available to achieve the gradual emission reduction standards should be accelerated and encouraged. These technologies include shoreside electricity, hybrid solutions, hydrogen, ammonia or synthetic fuels. A goal-based approach would give clear guidance to the shipping sector on the objectives to be reached while providing necessary flexibility for shipping, energy suppliers and ports on the choice of technologies allowing them to choose the most effective solutions. While ESPO is supportive of a policy framework that encourages investments in OPS and takes away the barriers for using OPS, it must be assessed case-by-case against other green solutions and must be seen in the context of the rapidly evolving zero-emission propulsion technologies (including hydrogen and ammonia).

The EU legislative proposals should increase the pressure on the IMO to roll out meaningful measures by 2023. ESPO believes that any European proposals such as an emission trading scheme (ETS), a levy or an innovation fund must be thoroughly examined in view of safeguarding the competitiveness of the EU port sector. In essence, a substantial part of the revenues from any market-based mechanism introduced must be used for port infrastructure investments and for supporting the use of clean fuel infrastructure. Environmentally differentiated port fees (incentive schemes) could be further adapted to the current challenges and encouraged. While streamlining between ports should be encouraged, the introduction, modalities of application and the level of potential environmental charges must remain a decision for each port managing body, taking into account the local situation and local environmental concerns and in accordance with the port's own roadmap.

11.7.2 EcoPorts Initiative and EcoPortsinSights Environmental Report

EcoPorts constitutes the main environmental initiative of the European port sector initiated in 1997 that has been fully integrated into the European Sea Ports Organisation (ESPO) since 2011. The founding principle of EcoPorts is to create a level playing field on environment through cooperation and sharing of knowledge between ports. EcoPorts provides two well-established tools to its members: the Self Diagnosis Method (SDM) and the Port Environmental Review System (PERS, certificate assessed by Lloyds register). The SDM procedure takes place in three steps as follows:

- SDM checklist: The SDM is a well-established and widely adopted, time- and cost-efficient methodology for identifying environmental risk and establishing priorities for action and compliance. Aggregated and anonymised data provided by EcoPorts members are used to build and update the sector's benchmark of performance in environmental management. The SDM is a concise checklist against which port managers can self-assess the environmental management programme of the port in relation to the performance of both the sector and international standards. The SDM checklist in particular addresses the fields of environmental policy –placing the focus on activities, aspects, objectives and targets – management organisation and personnel, environmental training, communication, operational management, emergency planning, monitoring, auditing and review. Individual port responses are treated confidentially.
- SDM Comparison: comparison of the port's SDM score with the European average.
- SDM Review: receive expert's advice and personalised recommendations.

Developed by ports themselves, the Port Environmental Review System (PERS) has firmly established its reputation as the only port sector-specific environmental management standard. The Port Environmental Review System (PERS) does not only incorporate the main general requirements of recognised environmental management standards (e.g. ISO 14001) but also takes into account the specificities of ports. PERS builds upon the policy recommendations of ESPO, while its implementation is independently reviewed by Lloyd's Register. Both of the aforementioned tools fit ports of different sizes and at different stages in the development of their environmental priorities.

The following ESPO (ESPO/EcoPorts 2019) Environmental Report – EcoPortsinSights – provides the latest trends of European sea ports concerning environmental issues. The data presented were obtained from 94 ESPO-member EU/EEA ports, which completed the online EcoPorts' Self Diagnosis Method (SDM). A set of environmental indicators were selected from the SDM to assess the environmental performance of EU ports. The SDM tool is also part of the EcoPorts pathway towards achievement of the port sector's own environmental standards, the EcoPorts' PERS.

11.7.2.1 Environmental Management Indicators

Table 11.1 presents the results of a set of selected environmental management indicators that are included in the EcoPorts' Self Diagnosis Method (SDM) providing information about the management efforts that influence the environmental

Table 11.1 Percentage of positive responses to the environmental management indicators. (Source: ESPO (2019))

	Indicators	2013	2016	2017	2018	2019	CHANGE 2013–2019
A	Existence of a Certified Environmental Management System –EMS (ISO, EMAS, PERS)	54	70	70	73	71	+17%
B	Existence of an Environmental Policy	90	92	97	96	95	+5%
C	Environmental Policy makes reference to ESPO's guideline documents	38	34	35	36	38	–
D	Existence of an inventory of relevant environmental legislation	90	90	93	97	96	+6%
E	Existence of an inventory of Significant Environmental Aspects (SEA)	84	89	93	93	89	+5%
F	Definition of objectives and targets for environmental improvement	84	89	93	93	90	+6%
G	Existence of an environmental training programme for port employees	66	55	68	58	53	−13%
H	Existence of an environmental monitoring programme	79	82	89	89	82	+3%
I	Environmental responsibilities of key personnel are documented	71	85	86	86	85	+14%
J	Publicly available environmental report	62	66	68	68	65	+3%

performance of a port, and it includes the percentage of positive responses to these indicators for the year 2019 as well as for 2013, 2016, 2017 and 2018 in order to analyse the variations over time.

Over the last years, the existence of an inventory of relevant environmental legislation has been the indicator with the higher percentage of positive responses demonstrating the awareness of ports about the requirement to comply with legislation. The indicator on the existence of an Environmental Policy (95%) follows in the second position, evincing ports' environmental commitment.

The definition of objectives and targets as well as the existence of an inventory of significant environmental aspects (SEA) are elements that are present in most of the ports (around 90%). These two indicators are the required first steps to start the implementation of any environmental management system (EMS). Related to this, the indicator on the existence of a certified environmental management system, i.e. ISO 14001, EcoPorts' PERS or EMAS has increased by 17% since 2013. Consequently, ports are not only willing to implement an environmental management system but also commit to comply with the standards in order to be certified.

Table 11.2 below demonstrates the number of ports that are certified with an internationally recognised environmental standard (environmental management system (EMS)). Out of the 71% of ports with a certified EMS, more than half have opted for ISO 14001 (53.7%) and almost one third of them for EcoPorts' PERS (26.9% – Table 11.2), making ISO and PERS the most popular standards in the port sector. Additionally, there are ports certified with more than one standard such as ports with ISO and EcoPorts' PERS (10.4%), followed by ports with all three certificates (4.5%) and ports certified with ISO and EMAS (3%). Another 1.5% of the ports is only certified with EMAS. Since 2013, the number of ports that are certified with EMS has significantly increased, manifesting the willingness of the sector to contribute to greening the supply chain.

Table 11.2 Breakdown of the EMS Certificates. (Source: ESPO (2019))

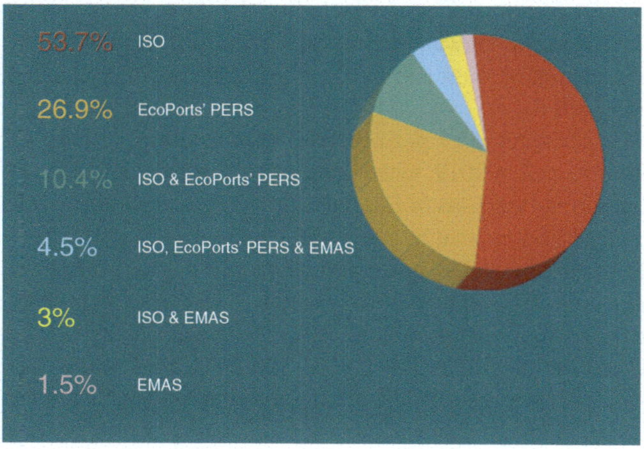

Table 11.3 Percentage of positive responses to environmental monitoring indicators. (Source: ESPO (2019))

Indicators	2013	2016	2017	2018	2019	CHANGE 2013–2019
Waste	67	79	88	84	79	+12
Energy consumption	65	73	80	80	76	+11
Water quality	56	70	75	76	71	+15
Water consumption	58	62	71	72	68	+10
Air quality	52	65	69	67	62	+10
Noise	52	57	64	68	57	+5
Sediment quality	56	63	65	58	54	-2
Carbon Footprint	48	47	49	47	49	+1
Marine ecosystems	35	36	44	40	40	+5
Terrestrial habitats	38	30	37	38	37	-1
Soil quality	42	44	48	38	32	-10

Importantly PERS, which is the EcoPorts' environmental standard and the only port sector-specific environmental standard available, is well recognised and preferred by the sector. EcoPorts' PERS is currently listed in a source of Good International Industry Practices (GIIP) in the World Bank Group Environmental, Health and Safety Guidelines for Ports, Harbors and Terminals and is recognised by several other port organisations and associations including the American Association of Port Authorities (AAPA), the Taiwan International Port Corporation (TIPC), the Port Management Association of West and Central Africa (PMAWCA) and the Arab Sea Ports Federation (ASPF).

Table 11.3 presents the percentages of positive responses listed in descending order, based on the results obtained in 2019, with regard to a set of indicators related to the Environmental Monitoring Programs of European ports. These results provide information on the percentage of ports that monitor selected environmental issues.

Since 2016, the three environmental issues regularly monitored by ports have remained the same. Following this trend, in 2019 waste was the most monitored indicator (79%), followed by energy consumption (76%) and water quality (71%). Water quality has increased the most over the last 6 years (+15%). Energy consumption, air quality and water consumption are monitoring issues that have

Table 11.4 Percentage of positive responses to indicators related to climate change. (Source: ESPO (2019))

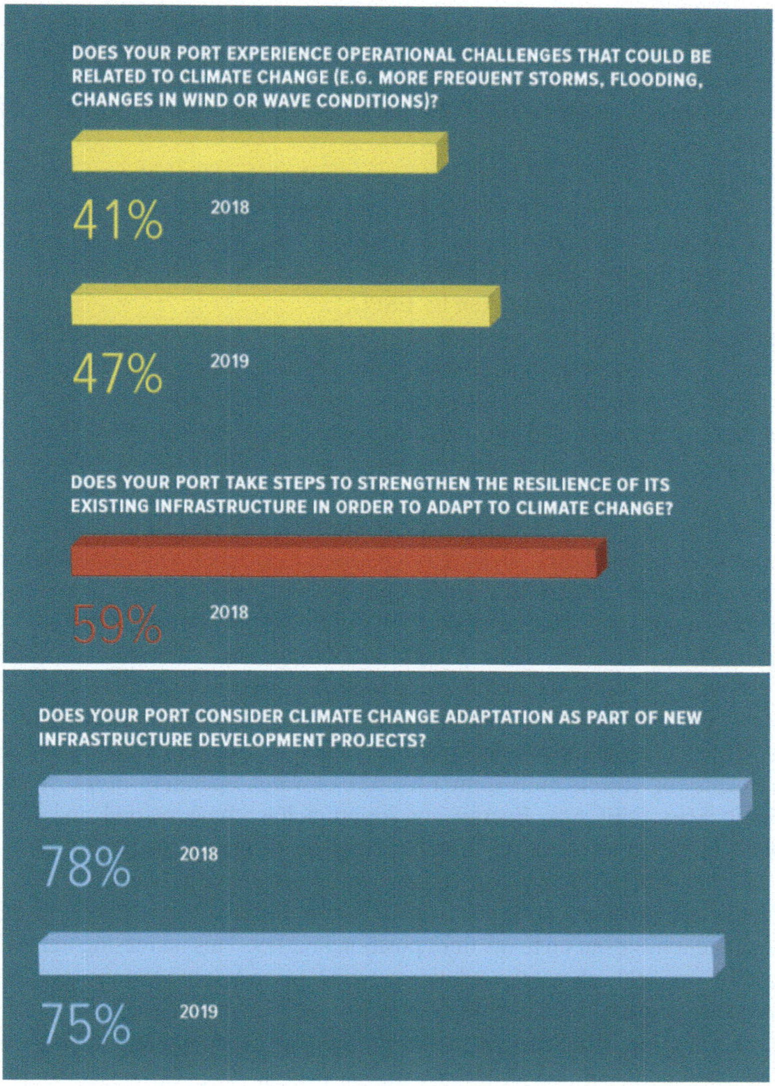

increased by around 10% since 2013. However, comparing the results with those of 2018, a reduction trend can be observed. Monitoring of soil quality has relatively decreased since 2017, though such monitoring is often associated with specific port development projects, carbon footprint monitoring has slightly increased since 2018.

Three new indicators related to climate change have been recently included in the report. The results are shown in Table 11.4. From 2018 to 2019 there has been an increase in the number of ports reporting operational challenges due to climate change from 41 to 47%. The same trend is observed with the percentage of ports that are taking steps to strengthen the resilience of their existing infrastructure to adapt themselves to climate change (62%). Finally, most of the ports are taking climate change into consideration for the development of their future infrastructure projects (75%). This is clear evidence that climate change and making infrastructure climate-proof is becoming a high priority. The latter are requirements of EcoPorts' PERS, ISO 14001 and EMAS.

Moreover, the 2019 EcoPorts Report presents the current issues faced by the port sector as top 10 environmental priorities of the European ports' managing bodies in 2019. These data are important as they identify the high-priority environmental issues in which port managing bodies are engaged and sets the framework for guidance and initiatives to be taken by ESPO.

The total set of Top 10 environmental priorities has been the same during 2017, 2018 and 2019. However, their relative positions have varied, with climate change rising from position ten to position three for instance. Air quality and energy consumption have occupied the first and second position since 2013 and 2016, respectively. These two environmental issues are of high relevance for European ports.

Air quality has been the first priority due to new legislation introduced over time. At the same time, air quality has increasingly been a priority for citizens of port cities and urban areas in general. Air quality has become a key determinant of public acceptance of port, and since more than 90% of European ports are urban ports, port managing bodies have this concern high on their agendas. In addition, EU regulations aiming to address air pollution include the implementation of the Sulphur Directive, the new National Emission Ceiling Directive, the introduction of the global 0.5% sulphur cap on marine fuels in 2020 and the IMO NOx Tier III requirements for vessels built from 1-1-2021 onwards operating in the North and the Baltic Seas (NECAs). Energy consumption has come second and has also remained in the same position. Improvement of efficiency, reduction of energy costs and the carbon footprint and climate change explain this stable position.

Climate change appeared in the Top 10 list for the first time in 2017 in the last position, and it has risen up to the third position in 2019. This increasing trend shows that complying with climate regulations, reducing carbon emissions and making infrastructure climate-proof are high priorities for European ports. In particular, many ports host industrial clusters in the port area and aim to organise their transition to a low carbon economy and become carbon neutral. Ship waste follows in sixth position and garbage/port waste in the seventh position. The implementation of the new EU Directive on Port Reception Facilities for ship waste will be among the priorities of ports for the next few years. This priority is also related to waste being the most monitored indicator for more than 5 years (see Table 11.3 above). Moreover, it evinces ports' readiness to contribute to addressing marine litter which is a great concern for local communities and civil society. Port development (land related) and water quality have decreased in priority, while dredging operations has remained in the same position. Dredging operations along with port development (land related) have been in the Top 10 rankings since 1996.

Table 11.5 Onshore power supply (OPS). (Source: ESPO (2019))

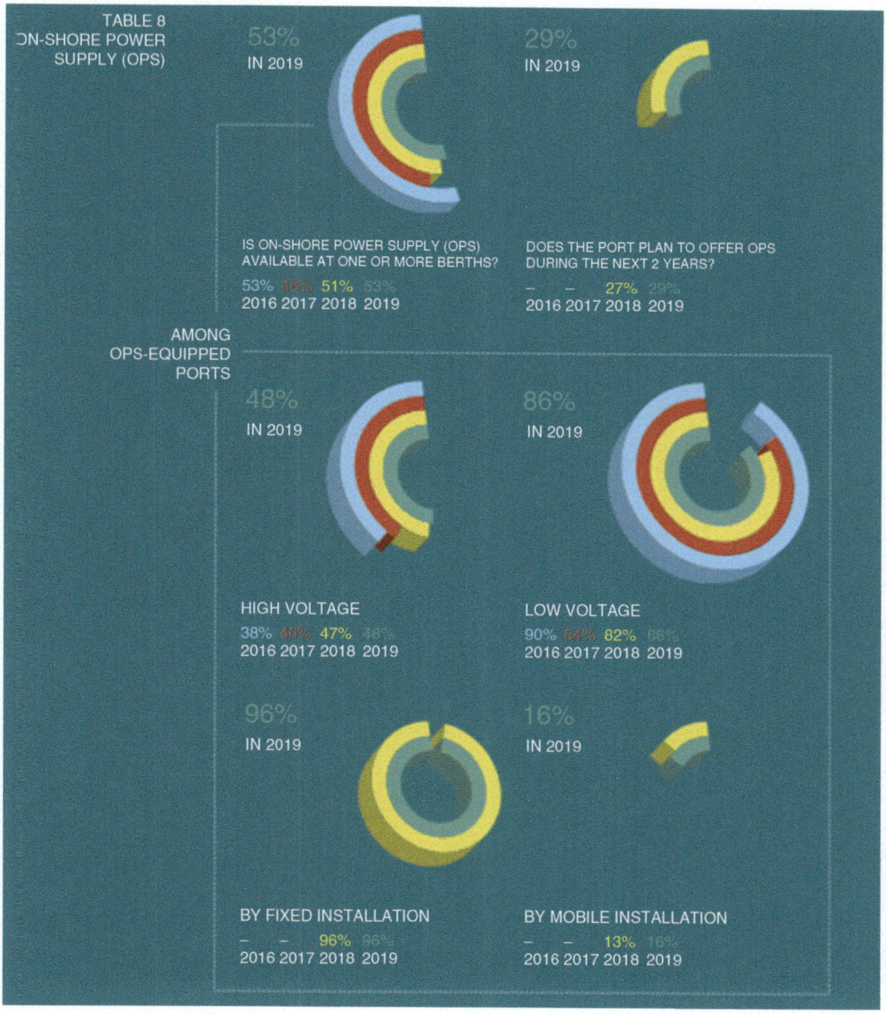

11.7.2.2 Green Services to Shipping

This section presents the share of EU ports that provide green services to shipping. It comprises three categories of indicators on the efforts made by the port managing bodies in order to contribute to greener shipping. These are the provision of onshore power supply (OPS), the provision of liquefied natural gas (LNG) bunkering facilities and environmentally differentiated port fees aiming to reward front-runners in the market and ships going beyond regulatory standards.

The EcoPorts SDM was updated in 2016 to enable the monitoring of the status and evolution of the green services that ports may choose to provide to their stakeholders. The results are benchmarked and presented in Tables 11.5, 11.6

Table 11.6 Liquefied natural gas (LNG). (Source: ESPO (2019))

and 11.7 and cover the period from 2016 until 2019. It should also be noted that the sample of the ports providing data for these three indicators was much smaller in the first year (2016) when the indicators were first introduced.

As shown in Table 11.5, more than half of the ports use OPS at their berths. In absolute figures, the ports offering OPS have increased from 32 (2016) to 50 ports (2019). Low-voltage OPS, with some exceptions, mainly relate to inland and domestic vessels as well as auxiliary vessels (e.g. tugs and/or other port authority vessels). In principle, the high-voltage OPS figure is more relevant for commercial seagoing vessels. The availability of high-voltage OPS has increased by 10% since 2016. In 96% of the OPS equipped ports, electricity is provided through fixed installations and in 16% of them through mobile installations. It should be noted

Table 11.7 Differentiated fees for 'greener vessels'. (Source: ESPO (2019))

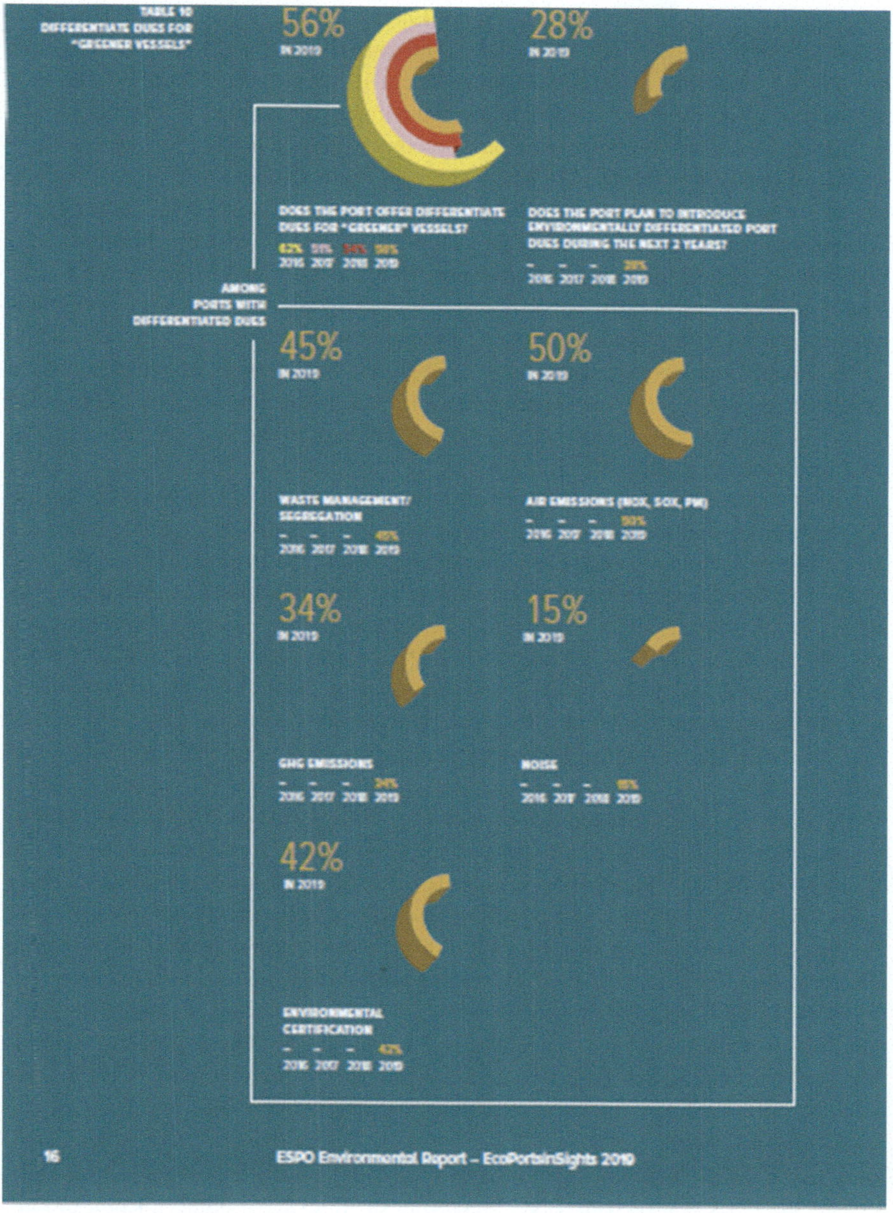

that some ports opt for both fixed and mobile installations. Interestingly, 29% of the ports seem to be planning to provide OPS in the next couple of years.

These results offer encouraging perspectives for the particular measure. However, the price differential between electricity and marine fuel and increased investment costs are the most significant barriers for the uptake of OPS. A recent evaluation paper of the European Commission on the Energy Taxation Directive (ETD) identified the problematic situation on OPS and recognised that 'the ETD does not provide for EU-wide preferential tax treatment of shore-side electricity and as a result, shore-side electricity is disadvantaged compared to onboard generation'.

Currently, electricity produced from the combustion of marine fuel on board of ships is tax exempt. However, when ships at berth connect with the shoreside electricity system, they are obliged to pay the energy tax applied to electricity. A limited number of EU Member States such as Sweden, Germany, Denmark and Spain have applied for and have been provided a temporary permit by the EU to apply a reduced rate of taxation to shoreside electricity for ships. This tax exemption has a time limit though and is obtained through a long administrative process at EU level. Taking into consideration these challenges, the Energy Taxation Directive should be reviewed to provide a permanent EU-wide tax exemption for OPS in order to be on equal terms with electricity generated onboard of the vessel which enjoys a tax exemption.

ESPO surveyed ports that currently provide OPS and found that levies applied to the electricity price are another significant barrier. Interestingly, in some cases the price differential remains high even after a tax exemption is provided by the EU, due to other national levies applied to the electricity price. In addition, technical challenges such as the frequency difference and additional investments for connection with the grid often prevent the uptake of OPS. In principle, ocean-going ships are 60 Hz equipped, and ports need to invest in frequency and high-voltage converters to address the frequency difference between the electricity from the grid (50 Hz) and the ship's equipment (60 Hz). Electricity shortage at city or regional level may be an additional barrier.

However, it has to be noted that investments in shoreside electricity remain high-risk investments since there is no guarantee or requirements whatsoever for the use of the available installations once provided. EU funding or co-funding of these investments by the users could contribute to sharing this risk. Policy measures on the port side such as the mandate for OPS under the Alternative Fuels Infrastructure Directive should be accompanied by corresponding measures for the port users.

Table 11.6 shows that the availability of LNG bunkering in the port sector continues to increase. This is a positive sign for the implementation of the Alternative Fuels Infrastructure Directive with regard to the provision by TEN-T core network ports of LNG bunkering facilities by 2025. Currently, one third of the ports offer this service to ships. This represents an increase of 10% since 2016. Interestingly, LNG is mainly provided by trucks (90%) and by barges (20%). Only 13% of the ports that provide LNG bunkering facilities have opted for non-mobile installation. It should be noted that some ports opt for more than one type of bunkering facilities while 24% of the ports mentioned the existence of ongoing projects to install LNG bunkering. This indicator was only added in 2018; hence, there is no data for the previous years.

As indicated in Table 11.7, environmentally differentiated port fees for ships that go beyond regulatory standards are set up in 56% of the ports. ESPO has been promoting this type of initiatives in its Green Guide (2012). However, it should be noted that, in principle, port fees make up a small part of the total port costs for ships and even smaller part of the total cost of a ship's journey. Thus, they do not aim to change investment decisions of shipowners but rather to reward and enhance the market reputation of the front-runners contributing to the greening of the supply chain as a whole.

Evidently, half of the ports that provide green discounts aim to encourage the reduction of air emissions, 45% of them to encourage better waste management and another 34% to encourage the reduction of GHG emissions. Environmental certification of ships is rewarded by 42% of them. Furthermore, 28% of them are planning to introduce environmentally differentiated port dues over the next years.

11.8 Conclusions

European Union has put into force a number of directives and regulations aiming to incentivise port and shipping companies to commit to comply with environmental standards. The European Green Deal, the most ambitious action plan of European Union, aims at increasing the EU's greenhouse gas emission reductions target for 2030 to at least 50% compared with 1990 levels, creating the most ambitious package of measures, accompanied with an initial roadmap of key policies in cutting-edge research and innovation, in green technologies and sustainable solutions. Most of the EU ports are actively working to protect the environment with the aim of achieving sustainable development. There has been a positive evolution of most of the environmental indicators since 2013. In principle, EU ports continue to improve their environmental performance and to maintain or even further enhance the declared policies of compliance, environmental protection and sustainable development. While it is difficult to identify and implement at once 'best practices' for all the environmental impacts that port activities generate, positive steps towards sustainable development and management are increasingly taking place.

References

Acciaro, M., Ghiara, H., & Cusano, M. I. (2014). Energy management in seaports: A new role for port authorities. *Energy Policy, 71*, 4–12. https://www.sciencedirect.com/science/article/abs/pii/S0301421514002262.

Asariotis R., Benamara, H., & Mohos-Naray, V. (2018). UNCTAD Research Paper No. 18 UNCTAD/SER.RP/2018/2018/Rev.1 United Nations Conference on Trade and Development (2018). https://unctad.org/en/PublicationsLibrary/ser-rp-2017d18_en.pdf

Bailey, D., & Solomon, G. (2004). Pollution prevention at ports: Clearing the air. *Environmental Impact Assessment Review, 24*(7–8), 749–774. https://www.sciencedirect.com/science/article/abs/pii/S0195925504000745.
Boerema, A., Van der Biest, K., & Meire, P. (2017). Towards sustainable port development. *Terra et Aqua, 149*, 5–17. https://www.researchgate.net/publication/321974575_Towards_sustainable_port_development.
Dabra, R. M., Ronza, A., Casal, J. A., Stojanovic, T. A., & Woolldridge, C. (2004). The self diagnosis method a new methodology to assess environmental management in seaports. *Marine Pollution Bulletin, 48*, 420–428.
Directive 2003/96/EC of 27 October 2003 restructuring the Community framework for the taxation of energy products and electricity. https://eur-lex.europa.eu/legal-content/EN/TXT/?uri=celex%3A32003L0096
Directive 2008/50/EC of the European Parliament and of the Council of 21 May 2008 on ambient air quality and cleaner air for Europe. https://eur-lex.europa.eu/legal-content/en/ALL/?uri=CELEX%3A32008L0050
Directive 2008/98/EC on waste (Waste Framework Directive). https://ec.europa.eu/environment/waste/framework/
Directive 2009/147/EC of the European Parliament and of the Council of 30 November 2009 on the conservation of wild birds. https://eur-lex.europa.eu/legal-content/EN/TXT/?uri=CELEX:32009L0147
Directive (EU) 2014/94 of the European Parliament and of the Council OF 22 October 2014 on the deployment of alternative fuels infrastructure (the AFID) https://eur-lex.europa.eu/legal-content/EN/TXT/?uri=celex%3A32014L0094
Directive (EU) 2016/802 of the European Parliament and of the Council of 11 May 2016 relating to a reduction in the sulphur content of certain liquid fuels. https://eur-lex.europa.eu/legal-content/en/TXT/?uri=CELEX:32016L0802
Directive (EU) 2019/883 of the European Parliament and of the Council of 17 April 2019 on port reception facilities for the delivery of waste from ships, amending Directive 2010/65/EU and repealing Directive 2000/59/EC. https://eur-lex.europa.eu/eli/dir/2019/883/oj
EcoPorts Environmental Report 2019. https://ecoports.com/publications/environmental-report-2019
EIT Climate-KIC. (2020). Deep Demonstrations. https://www.climate-kic.org/programmes/deep-demonstrations/. Accessed on 12 April 2020.
Environmental Impacts of International Shipping: The Role of Ports OECD Publishing. (2011). https://read.oecd-ilibrary.org/environment/environmental-impacts-of-international-shipping_9789264097339-en#page35
ESPO. (2012). "ESPO Green Guide; Towards excellence in port environmental management and sustainability" was formally adopted by the Executive Committee of ESPO in June 2012. https://www.espo.be/publications/espo-green-guide-towards-excellence-in-port-enviro
ESPO. (2019). European Sea Ports Organization's (ESPO's) 2019 Environmental Report. https://www.espo.be/media/Environmental%20Report-2019%20FINAL.pdf
ESPO. (2020). European Sea Ports Organization's (ESPO's) roadmap to implement the European Green Deal objectives in ports, February 2020 https://www.espo.be/news/espo-publishes-its-position-paper-on-the-european-
European Commission. (2019). The European Green Deal, Brussels, 11.12.2019 COM (2019) 640 final Communication from the commission to the European Parliament, the European Council, the council, the European Economic and Social committee and the committee of the regions. https://ec.europa.eu/info/sites/info/files/european-green-deal-communication_en.pdf
GEF-UNDP-IMO GloMEEP Project and IAPH, 2018a: Port Emissions Toolkit, Guide No. 1, Assessment of port emissions. https://glomeep.imo.org/wp-content/uploads/2019/03/port-emissions-toolkit-g1-online_New.pdf
GEF-UNDP-IMO GloMEEP Project and IAPH, 2018b: Port Emissions Toolkit, Guide No. 2, Development of port emissions reduction strategies. https://glomeep.imo.org/wp-content/uploads/2019/03/port-emissions-toolkit-g2-online_New.pdf

Habitats Directive 92/43/EEC. (1992). https://ec.europa.eu/environment/nature/legislation/habitatsdirective/index_en.htm

IMO International Convention for the Prevention of Pollution from Ships. (MARPOL 1973/78) http://www.imo.org/en/About/Conventions/ListOfConventions/Pages/International-Convention-for-the-Prevention-of-Pollution-from-Ships-(MARPOL).aspx

IMO MARPOL Annex VI. (2010). Prevention of Air Pollution from Ships. http://www.imo.org/en/OurWork/Environment/PollutionPrevention/AirPollution/Pages/Air-Pollution.aspx

IMO. (2020). Sulphur 2020 – cutting sulphur oxide emissions. http://www.imo.org/en/MediaCentre/HotTopics/Pages/Sulphur-2020.aspx

OECD/ITF. (2018). Reducing Shipping GHG Emissions: Lessons from Port-based Incentives. https://www.itf-oecd.org/sites/default/files/docs/reducing-shipping-greenhouse-gas-emissions.pdf

Özispa, N., & Arabelen, G. (2018). Sustainability issues in ports: content analysis and review of the literature (1987- 2017). Volume 58 (2018) SHS Web Conf., 58 (2018) 01022 DOI:10.1051/SHSCONF/20185801022

Puig, M., Wooldridge, C., & Darbra, R. M. (2014). Identification and selection of environmental performance indicators for sustainable port development. *Marine Pollution Bulletin, 81*(1), 124–130.

Reducing Shipping GHG Emissions: Lessons from port-based incentives-© OECD/ITF 2018. https://www.itf-oecd.org/sites/default/files/docs/reducing-shipping-greenhouse-gas-emissions.pdf

Regulation (EC) No 1013/2006 of the European Parliament and of the Council of 14 June 2006 on shipments of waste. https://eur-lex.europa.eu/legal-content/EN/ALL/?uri=CELEX%3A32006R1013

Regulation (EC) No 1221/2009 of the European Parliament and of the Council of 25 November 2009 on the voluntary participation by organisations in a Community eco-management and audit scheme (EMAS). https://eur-lex.europa.eu/LexUriServ/LexUriServ.do?uri=OJ:L:2009:342:0001:0045:EN:PDF

Regulation (EU) No 1315/2013 of the European Parliament and of the Council of 11 December 2013 on Union guidelines for the development of the trans-European transport network and repealing Decision No 661/2010/EU Text with EEA relevance. https://eur-lex.europa.eu/legal-content/EN/TXT/?qid=1589449235290&uri=CELEX:32013R1315

Regulation (EU) 2018/1999 of the European Parliament and of the Council of 11 December 2018 on the Governance of the Energy Union and Climate Action, amending Regulations (EC) No 663/2009 and (EC) No 715/2009 of the European Parliament and of the Council, Directives 94/22/EC, 98/70/EC, 2009/31/EC, 2009/73/EC, 2010/31/EU, 2012/27/EU and 2013/30/EU of the European Parliament and of the Council, Council Directives 2009/119/EC and (EU) 2015/652 and repealing Regulation (EU) No 525/2013 of the European Parliament and of the Council https://eur-lex.europa.eu/legal-content/EN/TXT/?uri=uriserv:OJ.L_.2018.328.01.0001.01.ENG&toc=OJ:L:2018:328:FULL

UNCTAD. (2019). Review of maritime transport 2019, https://unctad.org/en/PublicationsLibrary/rmt2019_en.pdf

Chapter 12
Circular Economy in National Smart Specialisation Strategies: The Case of Greece

Lena Tsipouri, Phoebe Koundouri, Lydia Papadaki, and Maria D. Argyrou

Abstract Greece is lagging significantly behind the EU average in its transition to the circular economy (CE); it needs significant acceleration to catch up. Its main advantage is good research skills. Monitoring indicators, European Semester recommendations, fines by the Court of Justice and national/international NGO assessments leave no doubt for that. This chapter (This work has received funding from the European Union's European Institute of Innovation and Technology under grant agreement N° 190,836.) presents a mapping exercise aiming at assisting the Greek authorities in using its Smart Specialisation Strategy (SSS), thus facilitating and accelerating the transition of the country to the CE. The combination of these two EU priority strategies and policies, totally distinct in terms of timing and primary target, poses significant challenges in terms of methodology, prioritisation and project coordination. The main lesson drawn from the Greek exercise is that the CE transition can be accelerated and become profitable, while using the cross-referencing methodology of SSS and CE strategy goals adapted for the needs and competitive advantages of each country or region proves a very helpful tool in the endeavour to accelerate the passage from the linear to the circular economy.

Keywords Circular economy · Smart Specialisation Strategy · Policy recommendations

---

L. Tsipouri (✉) · M. D. Argyrou
National and Kapodistrian University of Athens, Athens, Greece

P. Koundouri
School of Economics and ReSEES Laboratory, Athens University of Economics and Business, Athens, Greece

United Nations Sustainable Development Solutions Network Europe, Paris, France

EIT Climate–KIC Hub Greece, ATHENA Research and Innovation Center, Athens, Greece

ICRE8: International Centre for Research on the Environment and the Economy, Athens, Greece

L. Papadaki
United Nations Sustainable Development Solutions Network Greece, Athens, Greece

EIT Climate–KIC Hub Greece, ATHENA Research and Innovation Center, Athens, Greece

P. Koundouri (ed.), *The Ocean of Tomorrow*, Environment & Policy 57,
https://doi.org/10.1007/978-3-030-56847-4_12

12.1 Introduction

This chapter is about the mapping exercise aiming at assisting the Greek authorities in using its Smart Specialisation Strategy (SSS), thus facilitating and accelerating the transition of the country to the circular economy (CE). It describes the process of a pilot, supported by the EIT Climate-KIC, which can be used as a model by other member states wishing to couple their own SSS and CE strategies. The methodology followed consisted of desk research, interviews and a stakeholder consultation workshop conducted in September 2019 at the Ministry of Energy and Environment.

The project investigated the possibility to obtain synergies from the coordination of two top priority European Union (EU) policies, namely, SSS as a major tool of regional development policy and the transition to the CE as a main concern of environment policy.

The Circular Economy Strategy For a long time, environmental policy in Greece was (unsuccessfully) focusing almost exclusively on waste management with few exceptions; CE projects were fragmented, often considered identical to material recycling. Following the EU legislation and the Communication of the Action Plan for the CE, the Ministry of Environment and Energy adopted the National Circular Economy Strategy. Suggesting a methodology for refining and implementing the 2019 Strategy was the goal of the project described in this chapter.

The Smart Specialisation Strategy The adoption of a SSSs was an ex ante conditionality for the first time in the 2014–2020 programming period. Designing such a strategy created some unrest and a concern that focusing on few areas for thematic objectives 1 and 2 might discourage investments in the non-prioritised sectors of the economy. In addition, the timing did not allow for a systematic coordination between SSS and Operational Programmes (O.P.s). Due to the severe economic crisis, the European Structural and Investment Funds (ESIF) were the main source of development funding. Nationally funded investments shrank to a minimum because of the need to generate budget surpluses.

The combination of two totally distinct strategies and policies both in terms of timing and primary target poses significant challenges in terms of methodology, prioritisation and project coordination.

The rest of the chapter is structured as follows: In Sect. 12.2, we look at the external influences that have pushed the country towards the CE, their influence and potential incentives. In Sect. 12.3, we discuss the Greek context in more detail, looking at the indicators characterising the relative position of Greece in the adoption of the CE, as well as the design and implementation of the SSS. We then, in Sect. 12.6, describe the methodology used to assess common elements of the two strategies in the past. This methodology can prove invaluable if used ex ante in the next programming period rather than ex post. In conclusion, summarising the lessons learned and venturing some recommendations are included in Sect. 12.8.

12.2 External Influence

CE is mainstream in international organisations. The United Nations (UN) Sustainable Development Goals (SDGs) and the EU Circular Economy Action Plan constitute the most prominent efforts, promoting the CE influencing/supporting policy agendas in all their members (UN 2020; European Commission 2020). The UN SDGs are devised as a global, generic inspirational framework, whereas the EU regulatory framework is partly mandatory and partly discretionary with increasing incentives for its implementation creating obligations and opportunities for the member states.

12.2.1 *The United Nations Sustainable Development Goals (SDGs)*

The SDGs address the CE in the context of sustainability. Appendix 1 offers an overview of the CE-related content directly or indirectly included in the Sustainable Development Goals, as well as their respective targets and indicators. The distinctive feature of the SDGs is that, unlike the Millennium Development Goals, they address for both developed and developing countries. The SDGs are neither binding nor does the UN directly fund or otherwise support their integrated implementation but uses them as guidance for the developing countries' support by the various UN organisations, like UNCTAD, UNDP, etc. For the developed countries, they constitute an aid to national policies and are taken over by the OECD and the EU to be translated into more specific recommendations.

For Greece, which is a developed, yet middle-income country, four main lessons are derived for the design of its national CE strategy. The SDGs suggest regulatory interventions for practically all areas related to the CE with emphasis on the special treatment of hazardous waste, recycling/reuse of waste as well as sea and forest management. At least equally important to regulation are incentives for the private sector and civil society. The important role of technology for a profitable CE indicates that Research and Innovation (R&I) incentives can reinforce the role of the business sector and accelerate the transition. Several areas like wastewater treatment, renewable energy and energy efficiency and material consumption and production can benefit from CE-targeted R&I. While the primary sector plays a small role for GDP and employment in developed countries, its importance for the CE is disproportionately relevant: sustainable agriculture, supported by new technologies, precision agriculture and photonics, will contribute to the CE via sustainable food production. Competent public authorities, ministries or otherwise, are expected to join forces for introducing a CE strategy for agriculture. This is achievable in the short to medium term. Finally, specific additional tools from the public sector include green public procurement and monitoring of the carbon footprint and CE indicators, while from the private sector, CSR reporting will improve the business contribution.

12.2.2 The Circular Economy Transition in the EU

The EU has been very active early on in its vision for environmental protection and has integrated more aspects and policies in the introduction of the CE for its member states on December 2, 2015, when the European Commission put forward a package to support the EU's transition to a circular economy including an Action Plan with 54 specific actions (European Commission 2020). On March 4, 2019, the Commission informed on the complete execution of the Action Plan claiming that all 54 actions have been delivered or are being implemented. This is expected to not only protect the environment and generate sustainable growth but also create jobs, contribute to boost Europe's competitiveness and modernise its economy and industry. Hence, the influence of the EU CE strategy for the member states is multi-faceted: legally binding, inspirational and providing incentives.

The EU Action Plan for the Circular Economy outlines a set of both general and material-specific actions. While some obstacles to a circular economy are generic, different sectors and materials face specific challenges due to the particularities of the value chain.

General measures include product design, production process, consumption, from waste to resources (secondary raw materials), innovation, investment and other cross-cutting issues. While actions for specific materials and sectors include plastics, food value chain, critical raw materials, construction and demolition, biomass and bio-based products and review of fertilisers' legislation.

Many Directorates General (DG) of the European Commission, with a prominent role played by DG Environment, DG Grow, DG Research and Innovation and DG Energy, are directly or indirectly involved in the transition to the CE, using technical assistance, policy advice and financial incentives to support member states in their national policies. For the purposes of this chapter, we focus on the support provided by DG Regional Development, which codesigns the use of European Structural and Investment Funds (ESIF) with the member states and encourages them to use and to support the CE, using the following investment priorities[1]:

While the ESIF/SSS is an incentive for the CE in parallel with recommendations and encouragement, the European Commission uses the process of the European Semester to provide a framework for the coordination of economic policies across the European Union. It allows EU countries to discuss their economic and budget plans and monitor progress at specific times throughout the year. Each year, the Commission undertakes a detailed analysis of each country's plans for budget, macroeconomic and structural reforms. It then provides EU governments with

[1]The Partnership Agreement between the EU and the member states foresees the Operational Programmes to report based on specific thematic objectives and investment priorities, subject to the priorities decided in each member state.

country-specific recommendations for the next 12–18 months. The green economy is one of the themes, which includes a variety of environmental issues addressed in this context, though CE is not directly addressed yet.

12.3 The Greek Context

12.3.1 Snapshot of the Greek CE Performance

Compared to the EU average, Greece scores rather unsatisfactorily in its transition towards the CE. As demonstrated by Table 12.1, the country generates more municipal waste per capita or GDP with the exception of generation of waste excluding major mineral wastes per domestic material consumption, which is the only case it outperforms the EU average. It has a worse than average performance in all waste management with recycling rates being between 1/4 (in the case of biowaste) and close to the EU average (recovery rate of construction and demolition waste). The performance is at the order of magnitude of 1–10 in all secondary raw material indicators, while it is also underperforming in competitiveness and innovation (Table 12.2).

In a nutshell Greece is lagging significantly behind the EU average in its transition to the CE and needs significant acceleration to catch up.

The rather disappointing situation of the country concurs with the most recent European Semester Country-Specific Recommendations document for Greece (June 2019), where it is stated that "Treatment of solid waste and urban and industrial wastewater is the main area needing additional investment in order to align the

Table 12.1 Investment priorities potentially associated with the circular economy

6.1	Investing in the waste sector to meet the requirements of the Union's environmental acquis and to address needs, identified by the member states, for investment that goes beyond those requirements
6.2	Investing in the water sector to meet the requirements of the Union's environmental acquis and to address needs, identified by the member states, for investment that goes beyond those requirements
6.6	Promoting innovative technologies to improve environmental protection and resource efficiency in the waste sector, water sector and with regard to soil, or to reduce air pollution
6.7	Supporting industrial transition towards a resource-efficient economy, promoting green growth, eco-innovation and environmental performance management in the public and private sectors
7.3	Developing and improving environmentally friendly (including low-noise) and low-carbon transport systems
7.5	Improving energy efficiency and security of supply through the development of smart energy distribution, storage and transmission systems and through the integration of distributed generation from renewable sources

Table 12.2 Circular economy indicators

Indicator	Value	
	EU	Greece
Production and consumption		
1. EU self-sufficiency for raw materials (percentage)	36.4	N/A
2. Green public procurement	N/A	N/A
3. Waste generation		
Generation of municipal waste per capita *(kg* per capita*)*	486	504
Generation of waste excluding major mineral wastes per GDP unit (kg per thousand-euro, chain linked volumes (2010))	65	78
Generation of waste excluding major mineral wastes per domestic material consumption (percentage)	13.5	11.5
4. Food waste (million tonnes)	80	N/A
Waste management		
5. Recycling rates		
Recycling rate of municipal waste (percentage)	46.4	18.9
Recycling rate of all waste excluding major mineral waste (percentage)	57	N/A
6. Recycling/recovery for specific waste streams		
Recycling rate of overall packaging (percentage)	67.2	66.1
Recycling rate of plastic packaging (percentage)	42.4	38.2
Recycling rate of wooden packaging (percentage)	39.8	21.9
Recycling rate of e-waste (percentage)	41.2	34.2
Recycling of biowaste (kg per capita*)*	81	21
Recovery rate of construction and demolition waste (percentage)	89	88
Secondary raw materials		
7. Contribution of recycled materials to raw materials demand		
End-of-life recycling input rates (EOL-RIR) (percentage)	12.4	N/A
Circular material use rate (percentage)	11.7	1.3
8. Trade in recyclable raw materials (tonne)		
Imports from non-EU countries	5,905,135	536,071
Exports to non-EU countries	36,934,824	419,422
Intra EU trade	53,035,741	525,195
Competitiveness and innovation		
9. Private investment, jobs and gross value added related to circular economy sectors		
Gross investment in tangible goods (percentage of gross domestic product (GDP) at current prices)	0.12	0.04
Persons employed (percentage of total employment)	1.73	1.65
Value added at factor cost (percentage of gross domestic product (GDP) at current prices)	0.98	0.35
10. Number of patents related to recycling and secondary raw materials	338.17	0.5

Source: Eurostat (2019)

country's environmental protection standards with the rest of the EU. The management of solid waste continues to be a major structural challenge, with Greece still relying heavily on landfilling and mechanical-biological treatment instead of more modern techniques. In addition, the proportion of municipal waste that is recycled is only about a third of the EU average. Investments are also needed to improve water treatment, combat groundwater salinization, and support measures to prevent flooding and restore the natural flow of rivers" (European Commission 2019a, b, c). Moreover, the EU Court of Justice has imposed more than 100 million Euros of fines on Greece for non-compliance with EU law provisions in the fields of solid waste and urban wastewater treatment (European Commission 2019b; WWF 2019).

The EU is not the only one to express worries; WWF is systematically animadverting the country for its environmental performance, while this is confirmed by many national and international NGOs. The new exploration for oil in the Aegean Sea is one of the controversial issues for these organisations.

12.4 Policies and Governance for the CE

12.4.1 The Legal Landscape Before the Introduction of the CE Strategy

The predecessors to the CE policy in the country was waste management, and to a lesser extent, R&I supports measures for improvements in the energy and environment. The first law on recycling was adopted in 2001 (Law 2939/2001), but Greece has failed to achieve the targets it had set for itself on recyclables collection. This is attributed both to an inadequate mix of policies, to lack of incentives and to inadequate resources to the municipalities and citizens. The choices made were for very large and expensive recycling units with long delivery contracts processing large quantities of mixed waste. These options failed.

A most recent law adopted by the Parliament in 2017 (4496/2017) provides for sorting waste at the source, as well as ecological waste management. The aim of the government's policy in the programming period 2014–2020 was to harmonise Greek legislation with the European institutional framework so that by 2020 at least 70–80% of recyclable waste is collected at source. This objective was expected to be achieved at the municipal level with the participation of citizens, so that waste could be used as an important source for saving valuable and endangered raw materials.

This law was characterised by the introduction of a fee for plastic bags (which has since been introduced with spectacular impact on the reduction of plastic bags) and the introduction of specific measures to reduce the use of the plastic bag, in line with the provisions of Directive 2015/720/EU; the gradual achievement of new national targets and the reduction of waste resulting in landfill, below 30% by 2020; the creation of a National Public Information System and the upgrading of the recycling quality by requiring separate collection at source in at least four streams (containers) of packaging. Moreover, sorting at the source becomes mandatory in public spaces and utilities, operations in municipalities are optimising through incentives and disincentives, local management plans are formulated by the municipalities themselves, controls are intensified, and sanctions to stop producer avoidance to pay recycling fees are strengthened, while the Greek Recycling Organization (EOAN) is strengthened in human resources and organisational structure, and both citizens and municipalities are incentivised to participate in recycling.

The implementation was a disillusionment, mainly because of critical issues in this effort of modernising. There are frictions and opposing interests both at the different administrative levels (national, regional, local) as well as between the public and the private sector, while there have been significant regulatory omissions and missteps (because of the lack of regulation for recycling cooking oil, municipalities abandoned all efforts because they risked being treated as oil smugglers). Waste management projects generating revenue, fully or even partially commercial activities, require the control of competition rules and affecting the level of public funding. The control over the application of state aid rules to all operations has evolved into a deceleration factor (Mamalougkas 2019).

12.4.2 Policy Design and Implementation

Design and implementation for an encompassing environmental protection and energy policy is under the authority of the Ministry of Environment and Energy, while thematic ministries like the Ministry of Rural Development and Food, the Ministry of Shipping and the Aegean and the Ministry of Health take initiatives in the domain of their responsibilities. The Ministry of Development and Investment plays a decisive role in its role of designing and funding incentives for R&D as well as business investments. The major source of national funding comes from the ESIF, organised in Sectoral Operational Programmes and Regional Operational Programmes. In the latter case, an amount is foreseen for each region which is partly executed at the regional level and partly at municipal level. Hence, environmental missions are municipal, regional and national ESIF co-funded projects. Additional support is offered by the EU competitive calls (H2020, LIFE, COSME, European Territorial Cooperation Programmes, EIT KIC Greek Hub and NGO funded).

Policy implementation has until now not been sternly centrally monitored. This affects the ability to systematically collect the necessary data to construct pertinent indicators. Besides the centrally coordinated Operational Programme for Transport Infrastructure, the Environmental and Sustainable Development has many uncoordinated individual projects.

As pointed out in the European Semester recommendations, solid, at this stage, waste management is the most serious challenge for environmental protection and an opportunity for the transition to the CE. It remains heavily reliant on landfill (82% compared to 24% on average in the EU) and mechanical-biological treatment, as opposed to more modern techniques. Greece is at high risk of being unable to meet the EU's revised prevention and recycling targets (50% by 2020), as only 17% of municipal waste is currently recycled compared to an average of 46% in the EU. Despite declining in recent years, there are still some illegal landfills, resulting in costly infringement procedures for failing to comply with EU law on landfill and hazardous waste management.

However, progress has been made on the legal and institutional measures taken to increase the recycling of waste and to broaden EPR systems. The strategic framework for waste management is now being implemented with the approval of national and regional waste management plans. However, the use of financial means to incentivise prevention, reuse and recycling is inadequate, and existing systems appear to be lagging behind expected performance. Recycling has been gaining momentum but is still suffering from friction at the various administrative levels and the lack of a definitive and generally accepted governance structure, while production and consumption policies have not been a policy focus in the past. An initial mapping of actions includes:

(a) ***Research and innovation, GSRT*** [Gen. Sec. R&D] & NSRF Actions: 39 integrated research proposals for the 2016–2017 2-year period. There are two important actions in progress, namely, the *electronic platform of secondary materials at the Balkan level (INTERREG)* at the initiative of EDSNA [Association of Municipalities in the Attica Region – Solid Waste Management] and the participation of the Ministry of Environment and Energy and the *Environmental and Circular Economy Park of the Municipality of Heraklion* (UIA) at the initiative of ESDAK [Association of Solid Waste Management of Crete].
(b) ***An inter-ministerial committee on green public contracts***: It was established on June 13, 2017, in order to draft an Action Plan to promote green public contracts and submit proposals for planning a national policy within 18 months of its operation onset. The National Action Plan is approved by a Joint Ministerial Decision of the Minister of Economy and Development and the Minister of Environment and Energy. There is a similar proposal, prepared by a previous committee, for "greening" 18 product groups and a study proposal for a National Action Plan.
(c) **A mixed inter-ministerial working group titled "Industry Forum"**, established on February 2, 2016. The conclusions and proposals make an explicit reference proposal in favour of promoting circular economy in manufacturing through the "circular economy" model, which guides industrial entrepreneurship towards new productive operation models strongly characterised by innovation, environmental conservation and rational use of energy resources.
(d) **A mixed inter-ministerial working group titled "Agri-nutrition, Manufacturing, Tourism"** (September 16, 2016).
(e) **A mixed group of ELOT** [Hellenic Standardisation Organisation] Experts on "The Environment and Circular Economy" to effectively use international standards and to develop national standards concerning the environment, waste and circular economy, monitoring and participating in international and

Table 12.3 Number of projects funded by Sectoral Operational Programmes in 2014–2020

Operational Programmes	Number of projects	Total budget
Competitiveness, Entrepreneurship and Innovation	1	11,750,000
Transport Infrastructure, Environment and Sustainable Development	114	1,511,915,967
Total	115	1,523,665,967

European standardisation activities and recording domestic needs for models to help select standards of Greek interest (July 27, 2017).

(f) **An inter-ministerial group for the prevention of food waste and the creation of waste from food residues** (September 27, 2017).

(g) **A partnership on circular economy** (EU Urban Agenda), with the participation of six major urban centres (Oslo, The Hague, Prato, Porto, Kaunas and Flanders), four states (Finland, Poland, Slovenia, Greece), the European Commission (DG REGIO, ENV, CLIMA, RTD, GROW, etc.) and some organisations (CEMR, EUROCITIES, URBACT and EIB); the aim is the policies of circular economy in urban centres. The Greek working team includes participants from the Ministries of Environment and Energy, Shipping and Insular Policy, Tourism and the General Secretariat for Industry, under the coordination of the Ministry of Economy and Development (Special Service for Strategy, Planning and Evaluation – EYSSA).

The major sources of funding include the relevant Sectoral and Regional Operational Programmes, the Green Fund and the EU competitive programmes.

The relevant Sectoral Operational Programmes are grouped in two categories, namely, Competitiveness, Entrepreneurship and Innovation and Transport Infrastructure, Environment and Sustainable Development. Through a manual search in the integrated information system of the ESIF in the current programming period, 115 projects related to the CE (Thematic objectives 4, 6 and 7) have been funded with a budget of €1.5 billion until December 2019 (Table 12.3). The overwhelming majority of these projects are improving waste management.

Using the same methodology for the Regional O.P.s in the same investment priorities, a total of 89 projects were identified absorbing approximately €256 million in 6 years (Table 12.4). Again, the majority of these funds are earmarked for waste treatment.

The Green Fund (2020) operates in parallel with the O.P. Environment and Sustainable Development. The Green Fund may finance programmes drawn up by the Ministry of Environment and Energy or other ministries and their supervised agencies, decentralised administrations, local authorities, legal entities of the wider public sector, as defined by Article 1 of Law 1256/1982 and associations or other associations of legal and natural persons, which, in accordance with their statutory purposes, aim at the protection, upgrading and restoration of the environment. The Green Fund has a broader mandate but can be involved in CE actions.

Table 12.4 Number of projects funded by Regional Operational Programmes in 2014–2020

Region	Nr	€	%
Western Macedonia and Thrace	5	6,738,603	2.63
Attica	7	32,465,713	12.69
North Aegean	12	21,198,956	8.28
Western Greece	3	11,469,229	4.48
Western Macedonia	5	19,541,720	7.64
Epirus	12	68,666,395	26.83
Thessaly	6	21,362,330	8.35
Ionian Islands	3	3,189,799	1.25
Central Macedonia	1	49,600	0.02
Crete	35	71,212,199	27.83
Total	89	255,894,544	100.00

Lastly, the EU competitive programmes mentioned above (INTERREG, URBACT, H2020 etc.) can also support the funding of the CE implementation in Greece. An overview of the H2020 participations indicates that there are 83 Greek participations in bioenergy projects, 356 in biomass and 110 in various sustainability-related research projects, indicating high research skills.

The deficient implementation, divergent from the goals and the relative position of the country compared to the EU average, may be attributed mainly to the following bottlenecks and path-dependent, embedded deficiencies. In order to catch up, the Ministry of Environment and Energy designed a too ambitious to be implemented programme, while ministries have been following their own policy agendas with regions and municipalities designing their policies following major EU and national guidelines but with limited interregional interaction.

The business sector is not sufficiently sensitised and involved.[2] This is attributed both to the overall problems faced by private investment in the country, in particular during the 10-year long financial crisis,[3] and also to the complexities and long-term nature of profitability of investments in the areas of waste management, recycling and energy/materials efficiency. This has led to tensions between the private and the public sector and resistance to change. An additional issue undermining the profitability of aluminium and glass recycling is that Roma empty the recycle bins and steals the most valuable among streams diminishing the scale and profitability of private undertakings.

[2]This is a generic statement; few success stories exist but are small scale for the time being.

[3]Greece ranks 79th in the World Bank list of Ease of Doing Business https://openknowledge.worldbank.org/bitstream/handle/10986/32436/9781464814402.pdf and 59th in the Global Competitiveness report. http://www3.weforum.org/docs/WEF_TheGlobalCompetitivenessReport2019.pdf

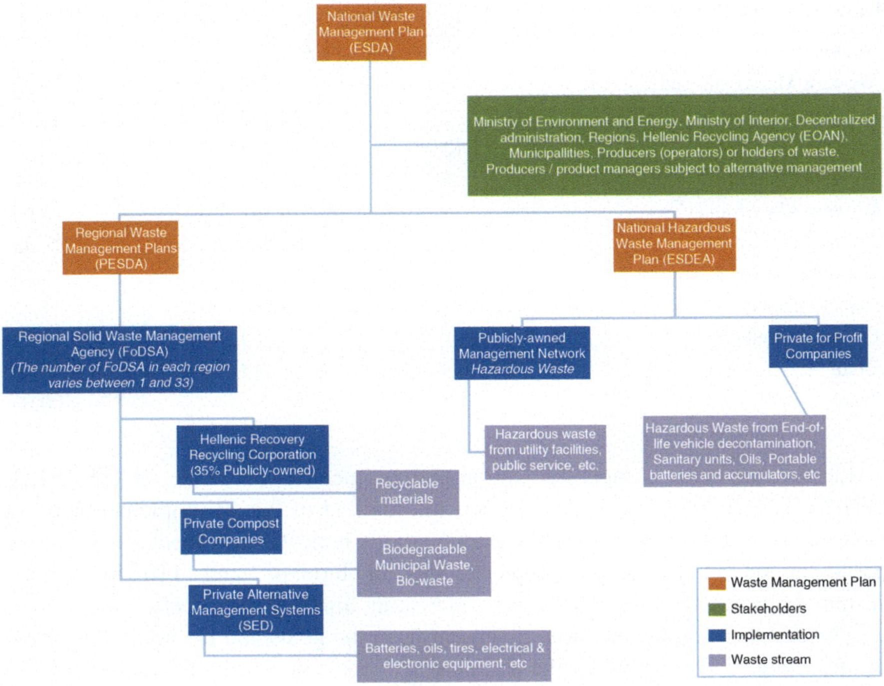

Fig. 12.1 Waste management organisations in Greece

12.4.3 Governance

The interaction between the national authorities, the regional authorities and the municipal authorities has developed with fragmented actions over the years and is rather complex (see Fig. 12.1) and bureaucratic. The governance structure is determined at the central level by the Ministry of Environment and Energy, which adopted the National Plan for Waste Management (NPWM), conceived with the aim of separating the different streams of waste in order to comply with the EU guidelines and respect the 2025 and 2030 targets.

Four main actors are involved, namely, Organisations of Solid Waste Management, the National Organisation of Recycling, Central Collection Facilities of Recyclable Materials (such centres are not geographically bounded, so there are interregional cooperation schemes for that reason) and Systems of Alternative Management. In addition, there are few fragmented private or NGO initiatives for smaller streams, like coffee residuals, etc., which are not thoroughly registered or documented.

The Organisations of Solid Waste Management (OSWM) which are public or publicly owned (intermunicipal) limited liability companies. The most active among them are the ones from Western Macedonia and from Eastern Macedonia-Thrace, as well as the intermunicipal one from Heraklion Crete. Many OSWM face financial liquidity problems and are unable to cover their obligations. In an effort to rationalise the process, Law 4555/2018 foresees the demolition of the existing OSWM and the creation of one per prefecture (Lawpost 2019). However, due to administrative and financial problems, the reorganisation has not yet materialised.

The National Organisation of Recycling (EOAN) is the responsible body of the Ministry of Environment and Energy for the design and implementation of policies for the prevention and alternative management of packaging and other products (EOAN 2019). It is responsible for approving national alternative product management systems, as well as for monitoring Greece's progress in recycling.

The Systems of Alternative Management (SED) are private, profit-oriented, officially licenced enterprises collecting specific waste streams for recycling. Such systems include the large, generic Hellenic Recovery Recycling Corporation (HERRCO) collecting in the same blue bins the basic recyclable materials and smaller, specialised collection streams (batteries and accumulators, electrical and electronic equipment, packaging and packaging waste, end life cycle vehicles, excavation construction and demolition waste, used vehicle tyres and lubricating oil waste) (HERRCO 2019; EOAN 2019). State aid rules applied here in the past and have caused bureaucratic delays. These systems are now reluctant to any governance changes, because the SED and the reorganisation of OSWM address the same market.

Funding is organised in a top-down and bottom-up mix: The Ministry, at the central level, has adopted its unrealistically ambitious National Programme for Waste Management (NPWM). All 332 municipalities of the country[4] had to come up with local waste management plans (LWMP), which would align with the ambitious targets of the NPWM. An indicative target set centrally was that 60% of biowaste had to be forwarded for composting. This was unachievable within the time limits foreseen. Once the municipalities adopted their LWMP, the corresponding (higher level administration) regional authorities[5] aggregated their suggestions into Regional Waste Management Plans (RWMP).

Figure 12.2 presents an overview of the system based on the experience of the programming period 2014–2020. However, a major change occurred in 2019, with the introduction of the circular economy strategy described below.

12.4.4 The Greek National CE Strategy (NCES)

The Ministry of Environment and Energy is precipitating the CE transition recognising its potential value. Two Secretaries-General are appointed, one with a

[4]Municipalities carry the responsibility for waste management in their territory.

[5]Regional authorities carry the responsibility for the implementation of the ESIF co-funded regional O.P.

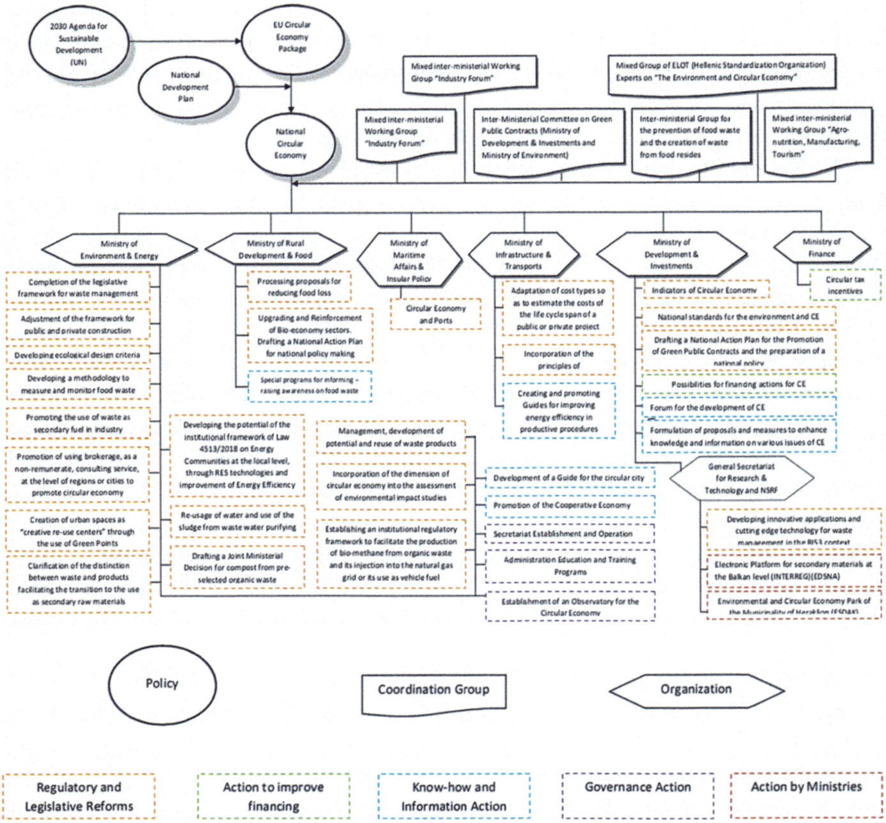

Fig. 12.2 The broader governance setup directly or indirectly involved with the CE

mandate to coordinate waste management and one for the natural environment and waters, carrying responsibility for the CE. A national strategy for CE was adopted in December 2018, after stakeholder consultation in an effort to accelerate circular economy actions and unlock growth potential. The strategy is composed of the following eight long-term goals (2030):

I. Integrating the criteria for ecological design/planning and analysis of product life cycle, avoiding the introduction of hazardous substances into their production and facilitating reparability and extension of product life span. The use of non-hazardous substances improves the quality of waste during the process of production, thus also reducing environmental income.
II. Effective implementation of prioritisation of waste management, promoting the prevention of creating waste and encouraging reusage and recycling.
III. Creating and promoting manuals for improving energy efficiency in procedures of production.
IV. Promotion of innovative forms of consumptions, such as the use of services instead of purchasing products or the use of electronic computers and digital platforms.

V. Promotion of a rational consumption model, based on information transparency in regard to the features of goods and services, their life span and energy efficiency.
VI. Facilitation and creation of appropriate channels for the exchange of information and the coordination between administrations, the scientific community and the economic and social agencies, so as to lead to synergies compatible with the transition to the circular model.
VII. Highlighting the significance of shifting from linear to circular economy, by promoting transparency in procedures, improving information given to citizens, training and raising social awareness.
VIII. Processing transparent and feasible indices for monitoring the implementation of the transition.

The public policy on circular economy focuses on the financing tools, planning and enactment of a regulatory framework and rules, as well as removal of bureaucratic obstacles, connection of small- and medium-sized entrepreneurship and social economy to technological innovation and the development and support of pilot/demonstration actions of circular economy and improvement of governance and networking and acceleration of relevant procedures (Ministry of Environment and Energy 2018).

Besides the sectors listed above, the spectrum of actions of implementation could be further enhanced by launching a series of institutional interventions that will reinforce circular economy, modular planning and open innovations; setting priorities on the basis of economic, social and environmental criteria; defining indicators to assess the circular economy model, and facilitating circular economy and industrial symbiosis entrepreneurial initiatives (administrative cost curtailing, public procurement premiums, eco-industrial parks, establishment of an appropriate regulatory framework and adjustment of the existing one). Smart financing tools with aids and tax reliefs together with utilising public investments, the NSRF, the investment bank, the Juncker package and other funds and resources could facilitate the implementation of the NCES.

More actions may include: enacting open licences, promotion of open technologies, utilisation of open innovation products, particularly in academic institutions and public administration; establishing specifications; creating databases and use of information for defining indicators to assess circular economy models in various sectors; incentives for developing social entrepreneurship, synergies and social economy in sectors of resource and material reuse (eco-industrial clusters, patent pools); policies facilitating the establishment of "smart factory" plants, which will be innovative, applying high technology, green, modular and probably digitised and a communicative strategy to raise citizens' awareness along with the provision of incentives.

Appendix 2 is presenting the Greek Action Plan of the Circular Economy gives a detailed timetable of the actions expected to be implemented within 2019. The Operational Action Plan envisaged regulatory and legislative reforms in a number of areas with a very optimistic attitude towards early implementation. Legal amendments are necessary to allow/facilitate measures, and they can be preceded by preparatory activities. It is important to ensure that wherever there are no mandatory regulatory provisions, regions will be allowed to proceed with their actions without expecting the national authorities to come up with recommendations.

Other areas requiring supporting actions include improving finance by investigating financing possibilities and circular tax incentives; know-how and information and governance actions. Some of the methods to educate citizens on CE are: a forum for the development of circular economy, a development of a guide for the circular city and promotion of the sharing economy and special programmes for informing raising awareness on food waste, promoting guides for improving energy efficiency in productive procedures and the formulation of proposals and measures to enhance knowledge and information on various issues of circular economy. Governance actions include the establishment and operation of a relevant secretariat, education and training programmes and the establishment of an observatory for the circular economy.

The NCES is significantly delayed but is an excellent list of topics to be discussed as basic themes for a future implementation plan. While the NCES is a significant step for awareness raising at political, policy and society levels, it should be viewed only as a good starting point: at this stage it constitutes a pertinent shopping list but is characterised with more enthusiasm than reality checks. It praises the CE and neglects its challenges. The conviction that the CE is beneficial for competitiveness relies on assumptions and contexts (like long-term investments, high profit margins and local manufacturing traditions) but neglects the significant bottlenecks of path dependence and finance in the country. Lagging regions, suffering from persistently low private investments and limited bank liquidity, tend to adopt short-term, survival solutions. Hence, a prerequisite for the NCES to succeed is a detailed context-specific analysis of cooperation, coordination and synergies to come up with solutions shifting from a short-termism behaviour to a realistic, profitable long-term strategy and the corresponding Action Plan.

12.5 The SSS Experience

Since the mid-1980s, the EU has adopted a cohesion policy whereby the less prosperous regions receive development aid from the European budget to make up for the uneven consequences of free trade following new economic geography and new trade theory insights. These transfers have been a major (occasionally, the only) funding source of development funding in Greece. The way policies were designed to absorb these funds has evolved over the years, as initially the funds were mainly spent on physical infrastructure and then gradually investments in a wider array of investment priorities to include technology, competitiveness and human capital.

In the programming period 2014–2020, the European Commission adopted for the first time the idea of Smart Specialisation Strategies as an ex ante conditionality for releasing the ESIF funds. Smart Specialisation is an innovative approach that aims to boost growth and jobs in Europe, by enabling each region to identify and develop its own competitive advantages. Through its partnership and bottom-up approach, Smart Specialisation brings together local authorities, academia, business spheres and the civil society, working for the implementation of long-term growth strategies supported by EU funds (European Commission 2019c).

Like all member states, the Greek authorities have designed SSSs both at a national level and in the 13 Greek regions, to allow them for selecting their own priorities. Ideally the SSS would be the rationale and background for the adoption of Sectoral and Regional O.P.s. However, because of institutional difficulties, the adoption of the SSS was delayed, and the O.P.s were adopted earlier, or in parallel, and adopted an extrapolating, path-dependent approach. Therefore, the SSS had a less decisive role than planned for. The Ministry and its sectoral O.P. had the primary role for the design, indicators and governance, leaving limited room for radical changes in case the SSSs had foreseen any.

The priorities selected by the regional SSS are presented on Table 12.5. The CE does not figure anywhere as the priorities are broader, but clearly the agri-food sector and energy saving, being indirectly associated with the CE, were part of practically all SSS, either as sectoral or as horizontal priorities. Within these priorities actions or projects adopting CE approaches could be included. SSS are rather broad and encompassing, the National SSS being the broadest and including eight sectors, practically reflecting the whole of the Greek economy (Table 12.5).

12.6 Linking the Smart Specialisation Strategy to the CE Transition: A Greek Pilot

After studying the CE transition progress and the SSS experiences in Greece, we focused on the main target of this study, which was to investigate the potential mutual reinforcement and synergies between the two. The methodology used was to systematically explore each one of the 14 SSS and complement the search with the relevant Sectoral Operational Programmes[6] trying to assess the extent to which their content corresponds to which NSCE goals.

In the 2014–2020 programming period, the crisis influenced the design of the Partnership Agreement with the EU towards favouring short-term projects with absorption targets and immediate, visible results inevitably neglecting longer-term investments. Environmental protection and the CE suffered under this approach, as they are by definition front-loaded in funding, but profitability only follows later. The adoption of the SSS and the corresponding Sectoral and Regional Operational Programmes could constitute an opportunity for Greece to embark into the CE transition with incentives for the business sector and knowledge-based investments.

Appendix 3 presents the results of the application of our methodology, namely, the Type of Intervention and Description by regional SSS and two sectoral O.P.s (O.P. – Competitiveness, Entrepreneurship, Innovation and O.P. – Transport Infrastructure, Environment and Sustainable Development), for all the cases, we consider

[6]Because of the lags and differences between the late adopted SSS and the important role of the sectoral O.P.s

Table 12.5 The Greek regional SSS priorities

	Priorities	Horizontal
Eastern Macedonia and Thrace	*Rural*, manufacturing, tourism (culture), emerging technologies (environment, energy, innovative building materials, hybrid technologies)	
Attica	Agri-food, design-intensive sectors, culture, media, tourism, Information & Communication Technologies, environmental technology, energy (*RES*, *energy saving*, smart grids), drug/health, intelligent and sustainable transport, shipbuilding	
North Aegean	*Agri-food sector development*, tourism, *Nature*, culture, innovation mechanisms and instruments, equal islands	
Western Greece	*Agricultural production*, aquaculture and food, tourism, culture, materials and microelectronics	ICT, energy
Western Macedonia	*Agri-food sector* with agri-livestock products, tourism sector, *waste management*, *energy and RES Heating*, fur sector	
Epirus	Primary sector, manufacturing, *agriculture*, gastronomy, industry experience, tourism, culture and the creative economy, Information & Communication Technologies, health and wellness, academic institutions and youth entrepreneurship	
Thessaly	Agri-food, creative tourism, *environment* energy, rehabilitation and advanced health services, metal and building materials	
Ionian Islands	Primary sector, *agri-food* and gastronomy, maritime economy, fisheries, aquaculture, marine tourism, industry of experience, tourism, culture and creative industry	
Central Macedonia	*Agri-food*, tourism, building materials, textiles and clothing	ICT, environment
Crete	*Agri-food complex*, cultural-tourist complex, *environmental complex*, knowledge complex	
South Aegean	*Agri-food*, fisheries and aquaculture, industry of experience, *green energy saving technologies*	
Peloponnese	*Agri-food sector*, tourism sector, Information & Communication Technologies, manufacturing and other dynamic sectors (materials)	*Energy*, *environment*, transport
Central Greece	*Agri-food*, experience industry, *green innovation*, *RES energy saving and production*, supporting the metal value chain	

potentially relevant for the CE. No evidence of explicit reference to the CE was found in the O. P., for agriculture. The outcome of this desk research was presented and discussed in the workshop held on September 2020 at the Greek Ministry and validated by stakeholders.

While the methodology proved interesting, there are two caveats we need to draw attention to at this pilot phase. As pointed out earlier, SSSs were adopted late in Greece, usually after the initial activities of the corresponding O.P.s were designed. Hence, whatever is included in the SSS design was not ipso facto translated into budgetary provisions. In the future the SSSs are expected to be closely linked, if not identical, with the O.P., and the methodology will prove more effective. In the current programming period, we focused on the SSS only, as this was the target of the study. Would we need a thorough study of the CE (unlinked to the SSS), we would need to differentiate between SSS and regional O.P.s. There are significant delays in the ESIF absorption and project implementation for most O.P.s and corresponding revisions. Consequently, CE actions suggested under the SSS may eventually not be implemented at all or at least not yet.

After extracting the relevant suggestions, we tried to match them to the individual goals of the CE. Appendix 4 presents the distribution of actions per region and NCES, leading to the following initial conclusions. Few activities suggested under the regional SSS address the CE directly. But many of the axes and interventions described per region and captured in Appendix 3, which are related to agricultural production, rationalisation of the economy, energy and the environment, may (or may not) be implemented in compliance with the NCES approach and principles, even though they were initially not adopted as such. The number and type of axes, interventions and related goals vary significantly across regions both in qualitative and in quantitative terms. Appendix 4 shows the relative frequencies per region and type of intervention. The highest number of CE-related interventions was found in Central Macedonia, followed by Central Greece and the Peloponnese. The lowest was in Western Greece and Western Macedonia.

The National SSS and Central Macedonia envisage interventions in all goals. The CE goals supported by the SSS are mainly goal 1 (Integrating the criteria for ecological design/planning and analysis of product life cycle); goal 2 (Effective implementation of prioritisation of waste management, promoting the prevention of creating waste and encouraging reusage and recycling); goal 3 (Creating and promoting manuals for improving energy efficiency in procedures of production) and goal 7 (Highlighting the significance of shifting from linear to circular economy) of the national strategy for CE. In particular, all regions (except for North Aegean) envisage activities addressing goal 2, followed by goal 7, followed by goal 1 and goal 3. Conversely, the lowest number of goals addressed is goal 6 by two regions only (Facilitation and creation of appropriate channels for the exchange of information and the coordination), followed by goal 8, addressed by three regions (Processing transparent and feasible indices for monitoring the implementation of the transition).

12.7 A Stakeholder Validation Workshop

A validation workshop on "Circular Economy Transition in Smart Specialization Strategy" was hosted by the Hellenic Ministry of Environment and Energy seeking the communication of the main outcomes of the project related to the synergies among the national strategy on circular economy and the research and innovation SSS and the discussion on the implementation of these strategies identifying the needs, barriers and strengths in Greece. Key stakeholders (e.g. Ministers, Region officials, Town mayors, entrepreneurs and start-ups) composed the audience in this participatory workshop, sharing their views on how the challenges of CE should be overcome and the opportunities to be exploited.

During the first part of the workshop, EIT Climate-KIC experts exercised system innovation tools to policymakers and other participants aiming to form a picture of their views on CE integration to S3. The stakeholders provided their opinion regarding CE implementation in S3 in Greece. The main opportunities and challenges identified during this process are discussed below.

12.7.1 Opportunities

In the class of the opportunities emerging from the workshop, first came the green growth of the Greek economy, which can be achieved through several interventions including the implementation of CE in different aspects. For example by redefining the regulations on recovered wastewater, new possibilities for the use of treated water will be created leading to an increase in wastewater use. The combination of a shift towards the growth of the primary sector, the strengthening of the IT market and the creation of specific IT brands will drive the above implementation goals, which in turn will generate more jobs.

There is a real growth scenario for Greece, where it is possible to demonstrate how an economic crisis can represent a moment of industrial transition to the circular economy. Cost deduction and sustainable consumption increase can be the right combination to overcome both economic and environmental critical issues. For example, an interaction between the agri-food and mining sectors could be created to reduce imports of raw materials and circularly manage existing mines. This would promote the creation of a new sector that of agri-mining. Equally required is the need for the circular economy to be technologically dressed, i.e. the use of deep tech to improve decision-making and the optimisation of all procedures. Besides that, it is possible to make the shipping repair industry a conservation industry.

The development of the green economy will be possible thanks to greater synergy between governmental bodies and through a public-private partnership. Cohesion in terms of economic development should be increased, and a differentiated approach to access more sustainable resources, especially in the extractive industry, should be ensured. This is the only way to increase competitiveness at international level. Due to the different perceptions of the concepts and expectations for the implementation of the circular economy by the different stakeholders, it is of great importance to start now for the design of the new programming period an intensive discussion on the national concept for the implementation of circular economy issues in any term that can help the sustainable development of the country.

In the preparation of the next programming period, the opportunity to incorporate the circular economy into the sectoral priorities of the new SSS should be included as a specific priority. As far as the public sector in the Central Greece region is concerned, it is necessary to apply the best practices necessary to assess the appropriateness of CE; create synergies and educational programmes and show problems to citizens on the decision-making side.

12.7.2 Challenges

There are also many challenges to be faced. These include achieving the objectives of the circular economy, creating measures and standards for companies to obtain the label as a "service or product of the circular economy" and developing the private sector more sustainably. Among the most difficult challenges is the regulation on the exploitation limits of raw materials, a traditional and sensitive sector with low sensitivity to innovation and a rigid legislative and operational framework. More interlocutors need to be found in different ministries to design an integrated approach for CE and SSS and create new synergies between the national and regional levels. A new policy and financing framework, strong stakeholder engagement and SME commitment are also needed. A general change of mindset accompanied by a general change in consumer behaviour is essential to make policy implementation possible.

A different perception of doing business should be disseminated, as well as a systematic approach based on motivation and not punishment. Other practical challenges that have emerged related to supporting the shipping recycling industry and deep tech (big data industries and blockchains) and limiting the phenomenon of greenwashes. Another theoretical challenge concerns the consumer economy, which implies the affirmation of the principle that to own is to be, thus pushing to maintain the ownership of values for individual things. In particular, it shows that the greatest problems in Western Greece and Attica are solid waste, recycling and sea plastic pollution.

12.8 Conclusions

The pilot EIT project described here was very timely for Greece, which is ready to adopt a revised SSS in view of the 2021–2027 programming period and needs to proceed with the revision of the adopted NCES.

The methodology proved operational and was positively received by the stakeholders involved in the September Workshop, despite some shortcomings deriving from the lack of experience and time constraints under which both the SSS and the NCES were designed in the past. In the new programming period, with the experience gained, the methodology coupling the two strategies can be mutually reinforcing and in particular help shift from a short- to a longer-term needed horizon, which is crucial for the CE success. The involvement of profitable activities by the business sector is an integral element to be incorporated in this codesign effort.

The project identified a series of problems in the Greek case. NCES focused primarily on waste management until 2019 and on areas where the country had a below EU average performance. Past efforts to gain ground were too ambitious to be implemented and led to disillusionments. The finally adopted NCES is more of a list of potential actions than a real, country-specific strategy. In an effort to sensitise stakeholders, it praises the CE and neglects to warn about challenges. The ambitious Action Plan could not be implemented within the timeframe foreseen, while the governance is not based on synergies, and private investments (a prerequisite for the

NCES to succeed) and profitability are not sufficiently involved. Finally, the multilevel/multi-actor interaction between national, regional and municipal authorities, the business sector and NGOs is rather complex and bureaucratic, and interests are often conflicting not complementary.

These problems are not insurmountable; if resolved and linked to the SSS, synergy opportunities can arise. Regional design based on competitive advantages can provide the long-term perspective and public-private cooperation the CE needs. Natural resources are available in the country and so are untapped secondary resources and waste. Using them as inputs for the revision of the SSS can lead to the generation of new competitive edges exploiting the scientific skills and expertise as well as productive tradition and know-how in technical trades. Agri-food is a sector with growth potential that requires modernisation and reduction of production costs. It is a priority in almost all SSS, so it is important to link it to its CE dimension in terms of production, consumption and waste management. A similar aspect can be exploited in the case of renewable energies.

Good governance and the exploitation of all available funding opportunities including good practices for new tools will be necessary to catch up with the EU average and even leapfrog. Policymakers should devise a generally acceptable coordination structure with clear demarcations of competences to ensure smooth cooperation between all administrative levels and the business sector while considering using new instruments, such as financial engineering and green or technology public procurement to enhance the role of business. Learning from the profitable CE investment in other countries and using EU peer learning opportunities in combination with pilots in Greek regions could enhance further the implementation of CE in Greece. Pilot the more mature Greek regions (Crete, Attica, Epirus in terms of R&D; Western Macedonia and Eastern Macedonia and Thrace in waste management), which, as demonstrated by this exercise, advance faster than others, so that the revised SSS in these regions could be used as pilots for the CE. Finally, the national strategy for circular economy needs to be redesigned in order to be more implementation-oriented. That is, include more explicit goals, set specific targets, propose a practical framework, and create a roadmap to enhance cooperation between the different levels of public administration, as well as to develop synergies across society and private and public sectors.

The revised SSS can be instrumental to help allocate more funds for projects promoting the circular economy in both national and regional level and by developing a modern strategy, which will incorporate the SDGs and EU Action Plan, and address upcoming challenges, while it will also transform the production model to become more sustainable and competitive in the long-term. Using the cross-referencing methodology of SSS and CE strategy goals adapted for the needs and competitive advantages of each country proves a very helpful tool in this endeavour. This is the main lesson drawn from the Greek exercise.

Appendices

Appendix 1: SDG Related to the CE

Goal	Target	Indicator
2. End hunger, achieve food security and improved nutrition, and promote sustainable agriculture	2.4 By 2030, ensure sustainable food production systems, and implement resilient agricultural practices that increase productivity and production, that help maintain ecosystems, that strengthen capacity for adaptation to climate change, extreme weather, drought, flooding and other disasters and that progressively improve land and soil quality	2.4.1 Proportion of agricultural area under productive and sustainable agriculture
6. Ensure availability and sustainable management of water and sanitation for all	6.3 By 2030, improve water quality by reducing pollution, eliminating dumping and minimising release of hazardous chemicals and materials, halving the proportion of untreated wastewater and substantially increasing recycling and safe reuse globally	6.3.1 Proportion of wastewater safely treated 6.3.2 Proportion of bodies of water with good ambient water quality
	6.6 By 2020, protect and restore water-related ecosystems, including mountains, forests, wetlands, rivers, aquifers and lakes	6.6.1 Change in the extent of water-related ecosystems over time
7. Ensure access to affordable, reliable, sustainable and modern energy for all	7.2 By 2030, increase substantially the share of renewable energy in the global energy mix	7.2.1 Renewable energy share in the total final energy consumption
	7.3 By 2030, double the global rate of improvement in energy efficiency	7.3.1 Energy intensity measured in terms of primary energy and GDP
8. Promote sustained, inclusive and sustainable economic growth, full and productive employment and decent work for all	8.4 Improve progressively, through 2030, global resource efficiency in consumption and production and endeavour to decouple economic growth from environmental degradation, in accordance with the 10-year framework of programmes on sustainable consumption and production, with developed countries taking the lead	8.4.1 Material footprint, material footprint per capita and material footprint per GDP 8.4.2 Domestic material consumption, domestic material consumption per capita and domestic material consumption per GDP

(continued)

Goal	Target	Indicator
9. Build resilient infrastructure, promote inclusive and sustainable industrialisation, and foster innovation	9.1 Develop quality, reliable, sustainable and resilient infrastructure, including regional and transborder infrastructure, to support economic development and human well-being, with a focus on affordable and equitable access for all	9.1.1 Proportion of the rural population who live within 2 km of an all-season road 9.1.2 Passenger and freight volumes, by mode of transport
	9.4 By 2030, upgrade infrastructure and retrofit industries to make them sustainable, with increased resource use efficiency and greater adoption of clean and environmentally sound technologies and industrial processes, with all countries taking action in accordance with their respective capabilities	9.4.1 CO_2 emission per unit of value added
11. Make cities and human settlements inclusive, safe, resilient and sustainable	11. 6 By 2030, reduce the adverse per capita environmental impact of cities, including by paying special attention to air quality and municipal and other waste management	11.6.1 Proportion of urban solid waste regularly collected and with adequate final discharge out of total urban solid waste generated, by cities 11.6.2 Annual mean levels of fine particulate matter (e.g. PM2.5 and PM10) in cities (population weighted)
	11.a Support positive economic, social and environmental links between urban, peri-urban and rural areas by strengthening national and regional development planning	11.a.1 Proportion of population living in cities that implement urban and regional development plans integrating population projections and resource needs, by size of city
	11.b By 2020, substantially increase the number of cities and human settlements adopting and implementing integrated policies and plans towards inclusion, resource efficiency, mitigation and adaptation to climate change, resilience to disasters, and develop and implement, in line with the Sendai Framework for Disaster Risk Reduction 2015–2030, holistic disaster risk management at all levels	11.b.1 Number of countries that adopt and implement national disaster risk reduction strategies in line with the Sendai Framework for Disaster Risk Reduction 2015–2030

(continued)

Goal	Target	Indicator
12. Ensure sustainable consumption and production patterns	12.1 Implement the 10-year framework of programmes on sustainable consumption and production, all countries taking action, with developed countries taking the lead, taking into account the development and capabilities of developing countries	12.1.1 Number of countries with sustainable consumption and production (SCP), national action plans or SCP mainstreamed as a priority or a target into national policies
	12.2 By 2030, achieve the sustainable management and efficient use of natural resources	12.2.1 Material footprint, material footprint per capita and material footprint per GDP 12.2.2 Domestic material consumption, domestic material consumption per capita and domestic material consumption per GDP
	12.3 By 2030, halve per capita global food waste at the retail and consumer levels, and reduce food losses along production and supply chains, including postharvest losses	12.3.1 (a) Food loss index and (b) food waste index
	12.4 By 2020, achieve the environmentally sound management of chemicals and all wastes throughout their life cycle, in accordance with agreed international frameworks, and significantly reduce their release to air, water and soil in order to minimise their adverse impacts on human health and the environment	12.4.1 Number of parties to international multilateral environmental agreements on hazardous waste, and other chemicals that meet their commitments and obligations in transmitting information as required by each relevant agreement
	12.5 By 2030, substantially reduce waste generation through prevention, reduction, recycling and reuse	12.5.1 National recycling rate, tons of material recycled
	12.6 Encourage companies, especially large and transnational companies, to adopt sustainable practices and to integrate sustainability information into their reporting cycle	12.6.1 Number of companies publishing sustainability reports
	12.7 Promote public procurement practices that are	12.7.1 Number of countries implementing sustainable

(continued)

Goal	Target	Indicator
	sustainable, in accordance with national policies and priorities	public procurement policies and action plans
	12.8 By 2030, ensure that people everywhere have the relevant information and awareness for sustainable development and lifestyles in harmony with nature	12.8.1 Extent to which (i) global citizenship education and (ii) education for sustainable development (including climate change education) are mainstreamed in(*a*) national education policies; (*b*) curricula; (*c*) teacher education and (*d*) student assessment
14. Conserve and sustainably use the oceans, seas and marine resources for sustainable development	14.1 By 2025, prevent and significantly reduce marine pollution of all kinds, in particular from land-based activities, including marine debris and nutrient pollution	14.1.1 Index of coastal eutrophication and floating plastic debris density
	14.2 By 2020, sustainably manage and protect marine and coastal ecosystems to avoid significant adverse impacts, including by strengthening their resilience, and take action for their restoration in order to achieve healthy and productive oceans	14.2.1 Proportion of national exclusive economic zones managed using ecosystem-based approaches
	14.3 Minimise and address the impacts of ocean acidification, including through enhanced scientific cooperation at all levels	14.3.1 Average marine acidity (pH) measured at agreed suite of representative sampling stations
	14.4 By 2020, effectively regulate harvesting and end overfishing, illegal, unreported and unregulated fishing and destructive fishing practices, and implement science-based management plans, in order to restore fish stocks in the shortest time feasible, at least to levels that can produce maximum sustainable yield as determined by their biological characteristics	14.4.1 Proportion of fish stocks within biologically sustainable levels
	14.6 By 2020, prohibit certain forms of fisheries subsidies which contribute to overcapacity and overfishing,	14.6.1 Degree of implementation of international instruments aiming to combat

(continued)

Goal	Target	Indicator
	eliminate subsidies that contribute to illegal, unreported and unregulated fishing, and refrain from introducing new such subsidies, recognising that appropriate and effective special and differential treatment for developing and least developed countries should be an integral part of the World Trade Organization fisheries subsidies negotiation	illegal, unreported and unregulated fishing
	14.a Increase scientific knowledge, develop research capacity, and transfer marine technology, taking into account the Intergovernmental Oceanographic Commission Criteria and Guidelines on the Transfer of Marine Technology, in order to improve ocean health and to enhance the contribution of marine biodiversity to the development of developing countries, in particular small island developing states and least developed countries	14.a.1 Proportion of total research budget allocated to research in the field of marine technology
15. Protect, restore and promote sustainable use of terrestrial ecosystems, sustainably manage forests, combat desertification, and halt and reverse land degradation and halt biodiversity loss	15.1 By 2020, ensure the conservation, restoration and sustainable use of terrestrial and inland freshwater ecosystems and their services, in particular forests, wetlands, mountains and drylands, in line with obligations under international agreements	15.1.1 Forest area as a proportion of total land area 15.1.2 Proportion of important sites for terrestrial and freshwater biodiversity that are covered by protected areas, by ecosystem type
	15.2 By 2020, promote the implementation of sustainable management of all types of forests, halt deforestation, restore degraded forests, and substantially increase afforestation and reforestation globally	15.2.1 Progress towards sustainable forest management
	15.3 By 2030, combat desertification, restore degraded land and soil, including land affected by desertification,	15.3.1 Proportion of land that is degraded over total land area

(continued)

Goal	Target	Indicator
	drought and floods, and strive to achieve a land degradation neutral world	
	15.4 By 2030, ensure the conservation of mountain ecosystems, including their biodiversity, in order to enhance their capacity to provide benefits that are essential for sustainable development	15.4.1 Coverage by protected areas of important sites for mountain biodiversity 15.4.2 Mountain Green Cover Index
	15.c Enhance global support for efforts to combat poaching and trafficking of protected species, including by increasing the capacity of local communities to pursue sustainable livelihood opportunities	15.c.1 Proportion of traded wildlife that was poached or illicitly trafficked
17. Strengthen the means of implementation and revitalise the Global Partnership for Sustainable Development	17.4 Assist developing countries in attaining long-term debt sustainability through coordinated policies aimed at fostering debt financing, debt relief and debt restructuring, as appropriate, and address the external debt of highly indebted poor countries to reduce debt distress	17.4.1 Debt service as a proportion of exports of goods and services
	17.7 Promote the development, transfer, dissemination and diffusion of environmentally sound technologies to developing countries on favourable terms, including on concessional and preferential terms, as mutually agreed	17.7.1 Total amount of approved funding for developing countries to promote the development, transfer, dissemination and diffusion of environmentally sound technologies

Appendix 2: The Greek Action Plan of the CE

Description	Time of implementation in 2019	Promoting – coordinating party
Waste management	1st half	Ministry of Environment and Energy (Ministry of the Interior, Ministry of Economy and Development)
Green public contracts, greening 18 product groups	1st half	Ministry of Economy and Development (Ministry of Environment and Energy, Ministry of Infrastructure and Transport, Ministry of the Interior)
Proposals for reducing food loss	2nd half	Ministry of Agriculture and Food (Ministry of Environment and Energy)
Construction projects framework	2nd half	Ministry of Environment and Energy (Ministry of Infrastructure and Transport)
Distinction between waste and products facilitating the transition to the use as secondary raw materials	1st half	Ministry of Environment and Energy (Ministry of Economy and Development, Ministry of the Interior)
Reusage of water and use of the sludge from wastewater purifying plants	2nd half	Ministry of Environment and Energy (Ministry of Economy and Development, Ministry of Agriculture and Food, Ministry of the Interior)
Developing innovative applications and cutting-edge technology for waste management in the RIS3 context	2nd half	General Secretariat for Research and Technology, Ministry of Economy and Development
Indicators of circular economy	1st half	Ministry of Economy and Development (Ministry of Environment and Energy, Ministry of the Interior)
Developing a methodology to measure and monitor food waste	1st half	Ministry of Environment and Energy (Ministry of Economy and Development, Ministry of the Interior)
Developing ecological design criteria	2nd half	Ministry of Environment and Energy (Ministry of Economy and Development and ELOT [Hellenic Standardisation Organisation], Ministry of Infrastructure and Transport)
National standards for the environment and circular economy	2nd half	Ministry of Economy and Development (ELOT [Hellenic Standardisation Organisation], Ministry of Environment and Energy, Ministry of Infrastructure and Transport, Ministry of the Interior)
Incorporation of the dimension of circular economy into the assessment of environmental impact studies	1st half	The Ministry of Environment and Energy in cooperation with the competent ministries at any given case: Ministry of Economy and

(continued)

Description	Time of implementation in 2019	Promoting – coordinating party
		Development (concerning entrepreneurial activities), Ministry of Infrastructure and Transport (concerning infrastructure), Ministry of the Interior (concerning licencing and municipal regulations)
Promotion of using brokerage, as a non-remunerated, consulting service, at the level of regions or cities to promote circular economy	2nd half	Ministry of Environment and Energy (Ministry of Economy and Development, Ministry of the Interior)
Creation of urban spaces as "creative resuse centres" through the use of Green Points/KAEDISP [Centre for recycling, training and sorting at source], turning them into "Green Centres"	1st half–2nd half	Ministry of Environment and Energy (Ministry of the Interior)
Promoting the use of waste as secondary fuel in industry	1st half	Ministry of Environment and Energy (Ministry of Economy and Development, Ministry of the Interior)
Establishing an institutional regulatory framework to facilitate the production of biomethane (green gas) from organic waste and its injection into the natural gas grid or its use as vehicle fuel	2nd half	Ministry of Environment and Energy (Ministry of the Interior
Drafting a Joint Ministerial Decision for compost from preselected organic waste	1st half	Ministry of Environment and Energy (Ministry of Economy and Development)
Upgrading and reinforcement of bio-economy sectors. Drafting a National Action Plan for national policy making	2nd half	Ministry of Agriculture and Food (Ministry of Environment and Energy)
Developing the potential of the institutional framework of Law 4513/2018 on energy communities at the local level, through RES technologies and improvement of energy efficiency	1st half	Ministry of Environment and Energy
Management, development of potential and reuse of waste products	2nd half	Ministry of Environment and Energy (Min. of Infrastructure and Transport)
Adaptation of cost types so as to estimate the costs of the life cycle span of a public or private project	2nd half	Ministry of Infrastructure and Transport
Incorporation of the principles of circular and sharing/cooperative economy in Sustainable Urban Mobility Plans (SVAK)	1st half	Ministry of Infrastructure and Transport (Ministry of Environment and Energy)

(continued)

Description	Time of implementation in 2019	Promoting – coordinating party
Circular economy and ports	2nd half	Ministry of Insularity and Island Policy (Ministry of Environment and Energy, Ministry of Infrastructure and Transport)

Appendix 3: O.P. and R.O.P. Interventions Possibly Linked to CE

Programme	Level	Type of intervention	Description
O.P. – Competitiveness, Entrepreneurship, Innovation	National	Action	Research, creation, innovation
O.P. – Competitiveness, Entrepreneurship, Innovation	National	Action	Enhancement of the environmental industry
O.P. – Competitiveness, Entrepreneurship, Innovation	National	Action	Green Point Network, development of separate waste collection systems and composting
O.P. – Competitiveness, Entrepreneurship, Innovation	National	Action	Open trade centres
O.P. – Competitiveness, Entrepreneurship, Innovation	National	Fund	Infrastructure
O P. – Transport Infrastructure, Environment and Sustainable Development	National	Priority axis	Priority Axis (14): CONSER VATION AND PROTECTION OF THE ENVIRONMENT – PROMOTING EFFICIENT USE OF RESOURCES
O.P. – Transport Infrastructure, Environment and Sustainable Development	Regional – Attica	Call	Integration and completion of integrated waste management infrastructure.
O.P. – Transport Infrastructure, Environment and Sustainable Development	Regional – Crete	Call	"Integration and completion of integrated waste management infrastructure".
O.P. – Transport Infrastructure, Environment and Sustainable Development	Regional – Epirus	Call	"Integration and completion of integrated waste management infrastructure".

(continued)

Programme	Level	Type of intervention	Description
O.P. – Transport Infra-structure, Environment and Sustainable Development	Regional – Ionian Islands	Call	Integrated municipal solid waste management actions in islands and small remote settlements in transition regions
O.P. – Transport Infra-structure, Environment and Sustainable Development	Regional – North Aegean	Call	Integrated municipal solid waste management actions in islands and small remote settlements in transition regions
O.P. – Transport Infra-structure, Environment and Sustainable Development	Regional – Peloponnese	Call	"Integration and completion of integrated waste management infrastructure".
RIS	National	Action	Increase investment in existing companies to introduce new products and services to the market and to develop and implement modern production methods
RIS	National	Action	Support businesses to build and expand advanced capabilities to develop new products and services in new areas
RIS	National	Target	Assist enterprises in the research and development of technologies for the collection, sorting, separation and exploitation of products derived from recyclable materials
RIS	National	Target	Development of technologies for the recovery, recycling and reuse of materials, development of alternatives for the absorption and economic recovery of materials recovered from special waste streams
RIS	National	Target	Development of innovative applications and cutting-edge technologies for the management of municipal waste (with a focus on biowaste), industrial waste and special waste streams, such as agri-food waste and tyres
RIS	National	Target	Produce high-quality environmental services to society to enhance transparency and mitigate social reactions, facilitating business involvement in the study and conservation of

(continued)

Programme	Level	Type of intervention	Description
			environmental resources and biodiversity. In this context, research and development of innovations in natural disaster planning, tackling the effects of climate change, exploiting genetic information on biodiversity, improving access to environmental information and involving businesses in conservation will be pursued of ecosystems and biodiversity
RIS	National	Target	An ecosystem-based approach to sustainable development through the creation of pilot research centres (e.g. upgrading laboratory equipment for the measurement of solid fuels, biofuels and secondary fuels from municipal waste), economic mapping of ecosystem services, etc.
RIS	Regional – Attica	Indicative actions	Products and processes for the management and exploitation of waste, trash and residues
RIS	Regional – Attica	Indicative actions	Trash and waste utilisation
RIS	Regional – Attica	Indicative actions	Products and processes for the management and exploitation of trash, residues and waste
RIS	Regional – Attica	Indicative actions	Development of innovative products and processes for the management and exploitation of waste, trash and residues for energy production and high value-added products
RIS	Regional – Attica	Indicative actions	Management and exploitation of waste, trash and residues for energy production and high value-added products
RIS	Regional – Attica	Indicative actions	Technologies and methods for reducing environmental footprint
RIS	Regional – Central Greece	Action	Modernising and applying sustainable farming methods
RIS	Regional – Central Greece	Action	Improvement of cover crops and introduction of hydroponics and aeroponic methods

(continued)

Programme	Level	Type of intervention	Description
RIS	Regional – Central Greece	Action	Certification, standardisation and introduction of innovations in the processing of agricultural and livestock products
RIS	Regional – Central Greece	Action	Support for new innovative manufacturing companies
RIS	Regional – Central Greece	Action	Development and introduction of innovations for the modernisation of farming methods and production protocols
RIS	Regional – Central Greece	Action	Use of green technologies in manufacturing and tourism
RIS	Regional – Central Greece	Action	Industrial coexistence programme to exploit waste and reduce resource use
RIS	Regional – Central Greece	Action	Small-scale investments in energy production in production units and holdings
RIS	Regional – Central Greece	Action	Documentation of the potential of biomass utilisation from various sources for energy production
RIS	Regional – Central Macedonia	Action	"Technological development projects to improve product quality (sustainability, eco-friendly)".
RIS	Regional – Central Macedonia	Action	"Synthesis of artificial marble using recyclable aggregates"
RIS	Regional – Central Macedonia	Action	"Manufacture of materials from renewable raw materials"
RIS	Regional – Central Macedonia	Action	"Water recycling in materials production processes"
RIS	Regional – Central Macedonia	Action	"Exploitation of by-product of fly ash from lignite combustion"
RIS	Regional – Central Macedonia	Action	"Utilisation of by-products and by-products – feed enrichment (bioactive foods)"
RIS	Regional – Central Macedonia	Action	"Utilisation of by-products and waste by biotechnological methods for the production of new products"

(continued)

Programme	Level	Type of intervention	Description
RIS	Regional – Central Macedonia	Action	"Knowledge platform in collaboration with operators and market"
RIS	Regional – Central Macedonia	Action	"Creation of permanent research – industry – consumer education and interconnection networks"
RIS	Regional – Central Macedonia	Priority	Reducing the environmental footprint of the agri-food processes
RIS	Regional – Central Macedonia	Priority	Reduce generation costs with emphasis on reducing energy consumption
RIS	Regional – Central Macedonia	Priority	Reducing the environmental impact of construction products and reducing their energy footprint (carbon footprint)
RIS	Regional – Central Macedonia	Priority	Smart buildings
RIS	Regional – Central Macedonia	Priority	Reduce generation costs with emphasis on reducing energy consumption (2)
RIS	Regional – Central Macedonia	Priority	Reducing environmental footprint – saving resources
RIS	Regional – Central Macedonia	Specific strategy	Specific strategy 2 (HS2) "empowering human capital in the direction of innovation – knowledge based on market needs".
RIS	Regional – Central Macedonia	Specific strategy	Specific strategy 3 (HS3) "emphasis on strategic areas of specialisation, utilisation of key enabling technologies/KETs and development of extroversion strategy".
RIS	Regional – Central Macedonia	Supporting strategy	"Strategies to support knowledge absorption and business dynamics". These include, inter alia, (a) lifelong learning activities in enterprises (high maturity); (b) awareness raising of businesses and stakeholders about the benefits and prospects of innovation, entrepreneurship-enhancing actions (average maturity) and (c) supporting demand for innovation through actions such as innovation vouchers (low maturity).

(continued)

Programme	Level	Type of intervention	Description
RIS	Regional – Central Macedonia	Supporting strategy	"Strategies to support recovery of lost soil in regions with high intensity in the primary sector." These include, inter alia, (a) regional offices for the promotion of entrepreneurship (high maturity) and (b) lifelong learning and skills development (high maturity)
RIS	Regional – Crete	Indicative implementation priorities	Precision agriculture in the country (climate and business organisation of production)
RIS	Regional – Crete	Indicative implementation priorities	Utilisation of agricultural waste products for the production of high nutritional value feed
RIS	Regional – Crete	Indicative implementation priorities	Develop protocols, reduce production costs, and improve the quality of cheese products in Crete
RIS	Regional – Crete	Indicative implementation priorities	Improving efficiency (reducing energy consumption of water systems, irrigation, wastewater management, solid waste management and generally large infrastructure)
RIS	Regional – Crete	Indicative implementation priorities	Development of technological applications to reduce the environmental footprint of economic activities (hotels, industries, hospitals and other public buildings)
RIS	Regional – Crete	Indicative implementation priorities	Pilot programme for the development and introduction of new technologies to reduce water losses
RIS	Regional – Crete	Indicative implementation priorities	Development of innovative municipal, industrial, livestock etc. solid waste management systems and pilot applications (prevention, collection, treatment, recovery/exploitation)
RIS	Regional – Crete	Indicative implementation priorities	Development of innovative municipal and/or industrial wastewater management systems and pilot applications (reuse, biofuel production, etc.)

(continued)

Programme	Level	Type of intervention	Description
RIS	Regional – Eastern Macedonia and Thrace	Action	Modernise the agri-food complex, and improve regional added value by using technologically driven innovation
RIS	Regional – Eastern Macedonia and Thrace	Action	Support for agri-food business investment plans for the introduction of RES technologies
RIS	Regional – Eastern Macedonia and Thrace	Action	Support business investment plans for the introduction of RES technologies
RIS	Regional – Eastern Macedonia and Thrace	Priority of intervention	Utilising modern production technologies and systems to reduce inputs into the production process
RIS	Regional – Eastern Macedonia and Thrace	Priority of intervention	Reduce the cost of production and disposal of products (including energy and transport)
RIS	Regional – Eastern Macedonia and Thrace	Priority of intervention	Utilising alternative uses of primary by-products, including their use as an energy resource
RIS	Regional – Eastern Macedonia and Thrace	Priority of intervention	Utilising technologies to reduce the volume and toxicity of waste along the value chain of the agri-food complex and further reduce its environmental footprint
RIS	Regional – Eastern Macedonia and Thrace	Priority of intervention	Rational management and utilisation of natural resources (water, agricultural land, forest wealth, pastures, etc.)
RIS	Regional – Epirus	Action	Development of applied research for food processing and by-product processing companies
RIS	Regional – Epirus	Action	Production of new innovative food products
RIS	Regional – Epirus	Action	Networking businesses that embody innovation
RIS	Regional – Epirus	Action	Improvement of existing farming methods
RIS	Regional – Epirus	Action	Utilising local potentials for fish production
RIS	Regional – Ionian Islands	Action	Production of agri-food products

(continued)

Programme	Level	Type of intervention	Description
RIS	Regional – Ionian Islands	Action	Use of green technologies in agricultural production
RIS	Regional – Ionian Islands	Action	Use of green technologies in the processing of agricultural products
RIS	Regional – Ionian Islands	Action	Development and use of green technologies in tourism
RIS	Regional – North Aegean	Project	Three pilot projects for waste management, treatment of waste mills, dairies, kernels for the purpose of creating new products
RIS	Regional – North Aegean	Project	Three pilot projects for the management of organic plant materials and waste for compost and/or pellet production
RIS	Regional – North Aegean	Project	Pilot project on green technology in accommodation or tourist service units
RIS	Regional – North Aegean	Action	Waste management
RIS	Regional – North Aegean	Action	Upgrading tourism offer business networking
RIS	Regional – Peloponnese	Area for intervention	Promoting precision agriculture
RIS	Regional – Peloponnese	Area for intervention	New technologies to promote and record water savings for irrigation
RIS	Regional – Peloponnese	Area for intervention	Developing innovative methods for the utilisation of waste, by-products and residues to reduce energy consumption and compost production (in collaboration with research institutes in the country)
RIS	Regional – Peloponnese	Axis	Development of tourism in harmony with the environment
RIS	Regional – Peloponnese	Specific target	Reducing environmental footprint, adaptation to climate change in the agri-food sector
RIS	Regional – South Aegean	Action	Modernising and applying sustainable farming methods
RIS	Regional – South Aegean	Action	Improvement of cover crops
RIS	Regional – South Aegean	Action	Introducing innovations in the processing of fish and aquaculture products

(continued)

Programme	Level	Type of intervention	Description
RIS	Regional – South Aegean	Action	Use of green technologies in agricultural production, processing and tourism
RIS	Regional – South Aegean	Action	Small-scale investments in energy production in production units and holdings
RIS	Regional – Thessaly	Area for intervention	Use of modern production technologies and systems to reduce inputs into the production process
RIS	Regional – Thessaly	Area for intervention	Reduce the cost of production and disposal of products (including energy and transport)
RIS	Regional – Thessaly	Area for intervention	Utilising alternative uses of primary sector by-products, including their use as an energy resource
RIS	Regional – Thessaly	Area for intervention	Implementation of innovative tools in the agri-food chain to reduce the volume and toxicity of their waste and further reduce their environmental footprint
RIS	Regional – Thessaly	Area for intervention	Reduce thermal energy costs by redesigning/modernising energy-efficient thermal processes and utilising biomass or waste while reducing the environmental footprint of the plants
RIS	Regional – Thessaly	Specific target	Support existing and new businesses to exploit patents and/or innovations, as well as support services to improve their productivity and/or to develop new products and services
RIS	Regional – Western Greece	Indicative actions	Development of innovative technologies for the protection and ecological restoration of water bodies (rivers, lakes, wetlands) in tourist areas and areas important for fisheries and aquaculture, etc.
RIS	Regional – Western Greece	Indicative actions	Development of materials recovery, recycling and reuse technologies

(continued)

Programme	Level	Type of intervention	Description
RIS	Regional – Western Greece	Indicative actions	Development of innovative applications and cutting-edge technologies for the management of biowaste and industrial waste and their energy utilisation especially in the agri-food sector
RIS	Regional – Western Macedonia	Indicative actions	Localised district heating systems with biomass utilisation
RIS	Regional – Western Macedonia	Indicative actions	Pilot waste refinery unit to optimise material sorting and align with the principles of industrial coexistence
RIS	Regional – Western Macedonia	Indicative actions	Development of Cluster Bioenergy and Environment (CLUBE) activities in Western Macedonia
RIS	Regional – Western Macedonia	Indicative actions	Exploitation of Western Macedonia's marine mining and quarrying by-products for the production of innovative/high value-added environmentally friendly materials
RIS	Regional – Western Macedonia	Indicative actions	Upgrade and expansion of biological cleaning (sludge compost management and safe disposal projects) of the Macedonian MABIK Meat Industry of Western Macedonia
RIS/EAFRD	Regional – Peloponnese	Action	Development of standard pasture management methods
RIS/EAFRD	Regional – Peloponnese	Action	Design, installation, operation of standard forage parks
RIS/EAFRD	Regional – Peloponnese	Action	Utilisation of by-products of dairies, olive mills with pilot application in demonstration units

Appendix 4: CE-Related Actions Per Region and NCES Goals

	Goal 1	Goal 2	Goal 3	Goal 4	Goal 5	Goal 6	Goal 7	Goal 8	No of interventions
Attica		6	2				6		7
Central Greece	3	4	2	1	1		4	2	9
Central Macedonia	8	9	4	2	2	2	2	1	20
Crete	1	5	3	1	1		3	1	9
Eastern Macedonia and Thrace	3	2	4				7		8
Epirus	2	2				1	3		6
Ionian Islands		2					4		5
North Aegean		4		3	3				6
Peloponnese	3	6	1				3		9
South Aegean	1	1	1				3		5
Thessaly	3	4	2				2		6
Western Greece	1	3	1	1	1		2		3
Western Macedonia		4	2				2		5
National RIS	7	3	2	3	1	1	1	1	7
National total	32	55	24	11	9	4	42	5	105

Number of interventions per region (5 in brackets)

	Goal 1	Goal 2	Goal 3	Goal 4	Goal 5	Goal 6	Goal 7	Goal 8	No of interventions
Attica		11%	8%				14%		7%
Central Greece	9%	7%	8%	9%	11%		10%	40%	9%
Central Macedonia	25%	16%	17%	18%	22%	50%	5%	20%	19%
Crete	3%	9%	13%	9%	11%		7%	20%	9%
Eastern Macedonia and Thrace	9%	4%	17%				17%		8%
Epirus	6%	4%				25%	7%		6%
Ionian Islands		4%					10%		5%
North Aegean		7%		27%	33%				6%
Peloponnese	9%	11%	4%				7%		9%
South Aegean	3%	2%	4%				7%		5%
Thessaly	9%	7%	8%				5%		6%
Western Greece	3%	5%	4%	9%	11%		5%		3%
Western Macedonia		7%	8%				5%		5%
National RIS	22%	5%	8%	27%	11%	25%	2%	20%	7%
National total	100%	100%	100%	100%	100%	100%	100%	100%	100%

Share of goal per region (%)

	Goal 1	Goal 2	Goal 3	Goal 4	Goal 5	Goal 6	Goal 7	Goal 8	Total
Attica		43%	14%				43%		100%
Central Greece	18%	24%	12%	6%	6%		24%	12%	100%
Central Macedonia	27%	30%	13%	7%	7%	7%	7%	3%	100%
Crete	7%	33%	20%	7%	7%		20%	7%	100%
Eastern Macedonia and Thrace	19%	13%	25%				44%		100%
Epirus	25%	25%				13%	38%		100%
Ionian Islands		33%					67%		100%
North Aegean		40%		30%	30%				100%
Peloponnese	23%	46%	8%				23%		100%
South Aegean	17%	17%	17%				50%		100%
Thessaly	27%	36%	18%				18%		100%
Western Greece	11%	33%	11%	11%	11%		22%		100%
Western Macedonia		50%	25%				25%		100%
National RIS	37%	16%	11%	16%	5%	5%	5%	5%	100%
National total	18%	30%	13%	6%	5%	2%	23%	3%	100%

Goals per region (%)

References

EOAN. (2019). National Organisation of Recycling. https://www.eoan.gr/en/. Accessed at 28 Sept 2019.

European Commission. (2019a). Council Recommendation. COM (2019) 508 final. https://ec.europa.eu/info/sites/info/files/file_import/2019-european-semester-country-specific-recommendation-commission-recommendation-greece_en.pdf

European Commission. (2019b). The EU Environmental Implementation Review 2019 Country Report – GREECE. Commission Working Paper, SWD (2019) 138 final. https://ec.europa.eu/environment/eir/pdf/report_el_en.pdf

European Commission. (2019c). Smart Specialisation. https://ec.europa.eu/regional_policy/sources/docgener/guides/smart_spec/strength_innov_regions_en.pdf

European Commission. (2020). EU Circular Economy Action Plan: A new Circular Economy Action Plan for a Cleaner and More Competitive Europe. https://ec.europa.eu/environment/circular-economy/. Accessed at 11 Aug 2019.

Eurostat. (2019). Monitoring Framework. https://ec.europa.eu/eurostat/web/circular-economy/indicators/monitoring-framework. Accessed at 19 Sept 2019.

HERRCO. (2019). Hellenic Recovery Recycling Corporation. http://www.herrco.gr/?lang=en

Lawpost. (2019). Law 4555/2018. https://www.lawspot.gr/nomikes-plirofories/nomothesia/nomos-4555-2018

Mamalougkas, N. (2019). Programming Period 2014–2020: Financing OP-TIESD. Critical Assessment and Implementation of IAS Projects. Ministry of Economy and Development, Hellenic Republic.

Ministry of Environment and Energy. (2018). National Circular Economy Strategy. Hellenic Republic. http://www.ypeka.gr/LinkClick.aspx?fileticket=pYSLQXgjjOU%3D&tabid=37&language=en-US

The Green Fund. (2020). http://www.prasinotameio.gr/index.php/en/. Accessed at 20 September 2019.

UN. (2020). SDG Indicators. Global indicator framework for the Sustainable Development Goals and targets of the 2030 Agenda for Sustainable Development https://unstats.un.org/sdgs/indicators/indicators-list/. Accessed at 11 Aug 2019.

WWF. (2019). WWF's 14th annual environmental law review in Greece. WWF Greece. https://www.wwf.gr/en/news/2203-wwf-s-14th-annual-environmental-law-review-in-greece

Chapter 13
Conclusions and Recommendations

Phoebe Koundouri and Lydia Papadaki

Abstract This chapter summarizes the concluding remarks and recommendations based on the analysis presented in the previous chapters. The chapters (This work has received funding from the European Union's seventh Framework Program under grant agreement N° 288,710 and N° 288,192, from the European Union's Horizon 2020 research and innovation program under grant agreement N° 675,680, N° 773,782, from the European Union's INTERREG Balkan-Mediterranean program under grant agreement MIS 5017160, from the European Union's European Institute of Innovation and Technology under grant agreement N° 190,880, N° 200,805, N° 201,166, N° 190,836, N° 190,744, N° 200,068 and N° 200,620.) of this book capture a wide spectrum of sustainable (e.g. economic, societal and environmental) challenges related to the seas presenting critical outcomes of marine and maritime research. The analysis in Chaps. 2, 3, 4, and 5 showed that MUOPs can potentially benefit from each other in terms of infrastructure, maintenance, etc. It is clear that the main sources of uncertainty about the viability of the projects are coming from the lack of precise knowledge on the operational conditions of the technology. In this context, MERMAID's assessment tool provided researchers with an intuitive way to evaluate multiple scenarios that would be hard and time-consuming to assess manually. Chapter 6 presents novel IT applications, which can facilitate producers to engage in the technology race, and Chap. 7 sheds light to the source-to-sea concept, which bridges the chasm for a better integration, cooperation and

---

P. Koundouri (✉)
School of Economics and ReSEES Laboratory, Athens University of Economics and Business, Athens, Greece

United Nations Sustainable Development Solutions Network Europe, Paris, France

EIT Climate–KIC Hub Greece, ATHENA Research and Innovation Center, Athens, Greece

ICRE8: International Centre for Research on the Environment and the Economy, Athens, Greece
e-mail: pkoundouri@aueb.gr

L. Papadaki
United Nations Sustainable Development Solutions Network Greece, Athens, Greece

EIT Climate–KIC Hub Greece, ATHENA Research and Innovation Center, Athens, Greece

P. Koundouri (ed.), *The Ocean of Tomorrow*, Environment & Policy 57,
https://doi.org/10.1007/978-3-030-56847-4_13

coordination of activities from the rural area until the ocean aiming at a harmonized and sustainable land-sea area. Chapter 8 focuses on marine research supporting that CES valuation can become an extremely useful tool that can bring to the surface the benefits derived from the cultural aspects of MPAs, while Chap. 9 depicts the key challenges of plastic marine litter. From the analysis carried out in Chaps. 10 and 11, it is clear that the maritime transport sector including ports not only are driving up global temperature but are essential part of the global economies. Ports role will be crucial in the law enforcement through reward schemes and priority entrance to ships complying with International and European regulation. Chapter 12 presents the circular economy approach, which can solve most of the challenges analysed in the previous chapters, and the synergies with the smart specialization strategies. All chapters underline the need for explicit targets and financial plans to be designed aiming at the implementation of ambitious climate and ocean-related targets.

Keywords Socio-economic methodology · Participatory approaches · Financial analysis · Web-based tool · Marine · Maritime · Sustainable development · Marine litter · Policy recommendations · Sustainable oceans

13.1 Introduction

The European Commission has initiated a program called Blue Growth, which is the EU's long-term strategy to support sustainable growth in the marine and maritime sectors. Within this context, several European-funded projects have been developed aiming to identify key components and useful answers to support the implementation of the European policies. This chapter summarizes the main findings of the analysis conducted in the previous chapters and highlights the directions and opportunities for future research under the auspices of the European Commission.

European FP7 projects called MERMAID (2015) and TROPOS (2012) have examined offshore platforms for multiuse of ocean space for energy extraction and aquaculture, along with their advantages in a floating platform and a multipurpose system. In Chap. 2 a socio-economic methodology for studying the social acceptance and the impacts of multiuse offshore developments is being analysed. This methodology is really important for developers and decision-makers to understand the views of local people and to learn about any potential concerns by assessing the social acceptance of the offshore developments. Chapter 3 reviews and discusses the participatory approaches employed in the MERMAID and TROPOS projects. The discussion draws on the methods employed in each case, the objectives and the obstacles encountered resulting in useful conclusions for participatory design. A common financial framework that permits to obtain comparable results of the financial performance of the different design concepts proposed in the Oceans of Tomorrow projects and demonstrated in the book "Oceans of Tomorrow" by Koundouri (2017) is defined in Chap. 4. The description of a decision support tool that was designed and developed for project MERMAID is the main objective of Chap. 5.

Chapter 6 explains several possible conceptual frameworks for the Cultural Ecosystem Services (CES) classification along with the monetary and nonmonetary methods for their valuation in the context of marine protected areas (MPAs) with application on AMAre (2020) and RECONNECT (2020). Chapter 7 presents the final outcomes of BlueBRIDGE (H2020) project drawing attention on the IT tools used (e.g. cloud computing infrastructure, virtual research environments) for the development of integrated production models. Rural and coastal sustainable growth and land-sea synergies are discussed thoroughly in Chap. 8, where COASTAL (H2020) project is presented. Chapter 9 covers marine plastic litter challenge and provides policy recommendations based on two EIT Climate-KIC projects: BL.EU. Climate (2020) and MEDFreeSUP (2020). Chapter 10 provides an overview of the marine transport industry, its role and relevance in sustainable development and the needed changes to be sustainable. Chapter 11 presents major challenges and opportunities related to ports and maritime transport presenting the EIT Climate-KIC (2020b) project, Deep Demonstration in Maritime Hubs.

13.2 Main Findings and Policy Recommendations

13.2.1 MERMAID and TROPOS Projects

The approach used in Chap. 2 is comprised a face-to-face survey with local people and tourists on Liuqiu island as well as in-depth interviews with particular stakeholders who are or will be potentially affected by offshore developments. Two methods are used, the first method in the mixed-method framework, applied to reach a wide range of participants by a multiple choice questionnaire survey. A large-scale survey was intended to give an indicative overview of how the offshore platform is perceived. The second method focused on the ability to explore in depth the range of perceptions of the platform and the potential impacts, that's why a qualitative study was designed. Interviews were used to obtain first-hand information on social realities as they are constructed and presented by different actors.

The methodologies used for assessing social acceptance in Chap. 2 are the following: (i) The Total Economic Value (TEV) is a standard theoretical approach used for capturing and describing the benefits derived from the different ecosystem services. The TEV is elicited from preferences of individuals using stated preference methods and revealed preference methods; the choice experiment (CE) method is included in this category. (ii) The social cost benefit analysis (SCBA) is a technique that assesses the monetary social costs and benefits of an investment project over a time period in comparison to a well-defined baseline alternative. Due to the project's expected long-run impacts on the local economy and ecology, its sustainability is to be tested using a long-run SCBA, and the net present value (NPV) of the project is to be estimated using different discount rate schemes. The NPV results reveal whether the net social benefits generated by the investment project of MUOPs are positive and significant well into the future.

The study in Chap. 2 found a general lack of awareness about the offshore platform project in Liuqiu Island. The results highlight the concerns on environmental effects of the platform and the unknown effects on existing industries, such as fishing and fishing processing. Moreover, there exist concerns from the respondents about who will be responsible for the platform construction and operation and from the residents about the possibility of the destruction of the platform by a storm wave. The findings also suggest that current public relation activities of the project are not proactive and point to the need to improve the involvement of local people and existing industries before the project can be carried out. The findings further show that cost and environment-related alternatives were factors that influenced to a high degree the stated preferences. The presence of energy and leisure facilities affected moderately the preference, and GDP effects and job creation were not deemed very important factors when the preference was stated. Local residents preferred not to install a platform and would rather focus on their traditional fishing ground, while tourists prefer the status quo the most, they would support the installation of the platform only if renewable energy and leisure facilities are provided.

Chapter 3 compares the participatory approaches employed at the TROPOS and the MERMAID projects that studied the development of multiuse platforms. The chapter describes the approaches and relates them to the objectives of participation with the aim to assess if participation was valuable. In this chapter, there are useful conclusions of participatory design, based on the findings of the two projects with the aim to inform future design processes.

In the TROPOS study, the concepts of Leisure Island Concept off Gran Canaria and the Green Blue Concept off Liuqiu Island Taiwan were used as examples. The results from Gran Canaria showed that there are concerns besides a general high acceptance of the Leisure Island among tourists as well as residents. Benefits for the tourist sector which are predicted to result in an increase of income and generation of jobs become confronted with various potential environmental impacts, in particular the disruption of marine species and habitats. The results from Liuqiu Island Taiwan also demonstrated a generally high acceptance of the Green and Blue platform among residents and tourists, although most participants had been unaware of the project. Despite the general acceptance of such a project, people also raised a number of specific concerns. These concerns are predominantly related to environmental impacts and unclear effects on local fishing and fish processing industries. Other issues that challenged the acceptance of the project include uncertain environmental impacts and adverse effects caused by the operation and construction of the platform. Perceptions of negative impacts were balanced against potential benefits for tourism, which is a crucial economic driver for Liuqiu Island.

The MERMAID project focused on four case study sites (North Sea, Mediterranean, Atlantic, Baltic) representative for European waters, each with local challenges. The results of the North Sea showed that the biggest challenge was to find solutions that could be profitable for all stakeholders, including risks and extra insurance costs. The results of the Mediterranean showed that a number of stakeholders initially opposed to the idea of including aquaculture farms in the multiuse platforms, because they were afraid of competition with the already existing coastal aquaculture. Despite this fear of competition, the design team decided not to limit the design by this argument, as this essentially was a plea for keeping a monopoly of the coastal aquaculture. In the Atlantic case, the final design included a combination of floating offshore wind turbines and wave energy generators. Stakeholders argued for the importance to select a site where conflicts with other interests are minimal. Multiuse platforms were considered to be able to provide revenues to both the local fishing community and local businesses. In the Baltic case, the eventual design combined wind turbines and offshore aquaculture by floating fish cages with trout/salmon production. This combination was interesting given the large-scale development of offshore wind and a technical risk assessment of the multiuse platforms appeared to be important. The stakeholders pointed out that there should be no negative effects on ecological conditions and that the artificial reefs on the wind turbine foundations should be protected as they have positive ecological effects.

From Chap. 3, it is clear that in the future, other projects should provide the necessary resources for creating an understanding of the locality as it is crucial for identifying the relevant stakeholders, their roles, objectives and resources. When eliciting stakeholders' view, selection bias should be avoided during both the preparation and interview stages. Thus, it is recommended to involve the relevant set of stakeholders for specific decisions. For example, in a technical scoping phase, it makes sense to only involve a small group of relevant experts, and in later project phases, stakeholders should be asked to pronounce themselves on few and reasonably well-defined design options that are possible for the specific offshore multiuse platform. Finally, this chapter demonstrated that shared knowledge and experience can contribute to more efficient and sustainable designs of offshore multiuse platforms. Acknowledging the stakeholders' perspectives enables surpassing potential obstacles and adjusts the design process is necessary. On the contrary, no dialogue or not considering stakeholders' point of view leads to risk of inefficient processes, the need to repeat procedures or even implement suboptimal solutions.

Chapter 4 presents the results of a comparative financial analysis performed to the three Oceans of Tomorrow projects. A homogeneous financial analysis of the Ocean of Tomorrow projects is developed, with the objective to test financial performance for all projects under the same assumptions and hypothesis, obtaining indicators that allow comparing the results between projects and comparing them themselves. In the same direction, a sensitivity analysis that enables the identification of the 'critical' variables of the project is carried out. Such variables are those whose variations, positive or negative, have the largest impact on the project's financial and/or economic performance. Moreover, a risk assessment of the projects is being calculated in this chapter.

This chapter has presented a transversal analysis of the outputs generated through different projects, trying to clarify the comparison among the existing alternatives and testing them from a standard economic and financial point of view. The results based on the comparison of different projects summarized in this chapter show a homogeneous ranking on the viability of the different alternatives and their business possibilities. The leadership in offshore activities is clearly located in renewable energy, and therefore the most promising combination proposals should be those where this industry appears. The results presented by aquaculture, seabed, logistics and leisure show more optimistic view in the related projects, but this optimistic analysis does not match with the real investments developed in those sectors in offshore areas. According to the results, the main sources of uncertainty about the viability of the projects are coming from the lack of precise knowledge on the operational conditions of the technology under untested conditions and from the intrinsic volatility of revenues to be obtained in the future with uncertainty in market conditions, environmental pressures and policy regulating measures. Finally, it is assumed that the initially recognized lack of maturity is still heavily restricting for further developments in offshore activities, and for now, it is unclear whether the shared approach creates expectations that will lead to a successful strategy.

In Chap. 5 an interdisciplinary web-based tool for applying socio-economic assessment in MERMAID project was described. This tool was part of the framework for assessing the socio-economic impact of MUOPs and, as such, utilized web and data analytics state-of-the-art technologies in order to provide researchers with a framework for evaluating feasibility and potential of each MUOP's proposed design and location. The assessment tool extends the standard process of financial analysis into an assessment that incorporates socio-economic, legal and technological environmental parameters. The tool provides the user with questionnaires for technical and legal feasibility assessment, as well as environmental impact assessment. It also materializes a streamlined robust methodology for the researchers and potential decision-makers.

The main results of the chapter are the following: (a) a data versioning system could be added into the tool, so as a researcher to be able to reproduce the analysis and compare the results; (b) the tool could potentially give the user the ability to preprocess raw climate data (wind, wave, currents, etc.) from existing repositories and shape them in the appropriate input format for the tool using graphical tools that would allow for a more sophisticated model and the expansion of the tool in other domains such as spatial analysis; (c) the enhanced tool could include a spatial data planning component that would use relevant project data (climate, engineering, ecological, socio-economic) to present the spatial alternatives, creating a suitability map; (d) the enhanced tool could visualize the outputs of sensitivity analysis; and finally (e) the described tool could perform manually cross validation to refine parameter selection for the simulation. From a general point of view, MERMAID's assessment tool provided researchers with an intuitive way to evaluate multiple scenarios that would be hard and time-consuming to assess manually. It managed to successfully capture a lot of information carefully chosen after multiple interactions with the stakeholders of the project MERMAID. This is a unique challenge given the multidisciplinary nature of the project, as well as the complexity introduced by the concurrent evaluation of different geographical sites.

13.2.2 Blue Growth and Source-to-Sea Sustainable Integration

Blue Growth, the long-term European strategy, aims to support sustainable growth in the marine and maritime sectors as a whole. Seas and oceans are drivers for the European economy and have great potential for innovation and growth. The strategy consists of three components: the development of sectors that have a high potential for sustainable jobs and growth, such as aquaculture, coastal tourism, marine biotechnology, ocean energy and seabed mining; the provision of knowledge, legal certainty and security in the blue economy; and sea basin strategies to ensure tailor-made measures and to foster cooperation between countries (including the Mediterranean Sea) (European Commission 2012). The following x chapters discuss blue growth under the spectrum of ecosystem services valuation, aquaculture and virtual research environments.

Chapter 6 presents BlueBRIDGE (H2020) project, which aimed at the enlargement of the spectrum of growth opportunities in distinctive Blue Growth areas, and specifically it provides a brief overview of the virtual research environments (VREs) developed within the BlueBRIDGE project with focus on aquaculture (BlueBRIDGE 2020). Aquaculture can be core to Blue Growth targets with benefits exceeding the private benefits. The efficient and sustainable management of aquaculture production, at micro- but also at macro-level, requires the use of cutting-edge technology, novel IT applications and integrated socio-economic tools. The tools and the methodology developed within the scope of the BlueBRIDGE project allow for the use of advanced IT applications and facilities and the introduction of the wider socio-economic and environmental effects of aquaculture into the production management. This can support well-informed management and decision-making. Developing and proposing tools and methods enable the estimation of an integrated value of production that looks beyond output maximization. Easy to use tools facilitate all producers and the sector overall to engage in the technology race and use it at their own benefit. From a policy perspective, the outputs of the project enable the well-informed and forward-looking decision-making and target setting.

In Chap. 7, COASTAL (H2020) framework is demonstrated, which aims to increase land-sea synergies via an innovative dual approach using systems dynamics coupled with participatory methods by involving local stakeholders from representative EU coastal regions to codesign business roads map and policy alternatives (COASTAL 2020). The source-to-sea continuum considers the land and the sea as a single component (e.g. from land to freshwater, delta, estuarine, coastline, near shore and ocean, connected through flows of waters, including sediment, pollutants, materials, biota and related ecosystem services). Coastal areas and marine water are inevitably impacted by land-based activities (agriculture, forestry, heavy industry and urbanization) due to unsustainable land use, soil degradation and pollution of inland freshwater.

Initial work of the COASTAL project has yielded preliminary findings which indicate that new, innovative practices such as combined activities, alternative forms of tourism and development of renewable energies offshore or coastal risk management strategies can play a role in co-creating a common vision for the development of an interconnected rural-coastal-sea region. The project aims at identifying main issues and business opportunities from representative case studies and the development of an inventory of best practice, successes and lessons learned regarding land-sea synergies and coastal-rural collaborations. Policies are still fragmented, but international initiatives (political and scientific) work towards unified, multi-sectoral framework and a better integration of the land and sea ecosystems. The source-to-sea concept highlights the gaps and needs for a better integration, cooperation and coordination of activities from the rural area until the ocean in order to achieve a harmonized, sustainable development of a land-sea area.

13.2.3 Marine Protected Areas

Chapter 8 presents possible conceptual frameworks for the cultural ecosystem services (CES) classification along with the monetary and nonmonetary (revealed and stated preference) methods for their valuation. The methodology used to operationally define CES in socio-economic models, in a way that they can be useful in policy agenda and in decision-making as well as classify is a literature review approach. Besides the lack of well-established and readily applicable definition, literature seems to agree that CES are *nonmaterial* and *contribute benefit to people in terms of well-being, mental and physical health.* An efficient context to achieve an improved management of ecosystem services (ES) are the marine protected areas (MPAs), which are directly aligned with ES taking under considering that in most marine areas, human activities are not spatially managed and monitored, while human impacts on ecosystems services are not taken into account when management initiatives are considered.

The study supports that CES valuation can become an extremely advantageous tool that can shed light to the benefits derived from the cultural aspects of MPAs, guiding policymakers and management authorities. Integrated and adaptive management will support MPA managers to identify and adopt policies and best practices that involve both cultural and natural resources at the ecosystem and landscape levels. So far, the variety of conceptual frameworks around the CES categorization has undermined this opportunity. This chapter justifies that stated preference methods have the lion's share in this debate, and there seems to be a consensus that the more CES will become important, the more these methods will need to be developed to accommodate the specificities associated with these services. Recent projects that apply the combination of monetary and nonmonetary valuation methods in MPA management are AMARE (2020) and RECONNECT (2020).

13.2.4 Marine Plastic Litter

Chapter 9 discusses the marine litter issue in the Mediterranean Sea through two EIT Climate-KIC (2020a) - funded projects BL.EU. Climate (Climate Innovation in Southern Waters), which is implemented in 2019 and MEDFreeSUP (tacking single-use plastic item uses in the Easter Mediterranean Sea), which is the follow-up project, and it will be implemented in the period 2020–2021. The methodology used by BL.EU. Climate (2020) project conducted an extensive stakeholder mapping in all case studies in Greece, Portugal and Croatia, validating through interviews and surveys targeting key challenge owners (e.g. fishermen and tourists); presenting the outcomes at different workshops conducted in each country seeking to trigger discussions among the participants on potential solutions to prevent, reduce and collect marine litter; focusing mostly on plastics; and designing a strategic roadmap by all three countries, identifying measures to reduce the negative effects caused by plastic waste in the future, supporting policymakers, industries, consumers and civil society to improve systems design, replacement, refuse, recycling and reuse of plastic.

The BL.EU. Climate study found that one of the greatest barriers in Greece is the knowledge gap across stakeholders. Most of the respondents understand that the natural environment is at crisis, and this is caused mainly by anthropogenic factors. However, the most affected by marine litter actors (e.g. fishermen and tourists) seem to lack a basic understanding of the marine plastic pollution and its detrimental effects on the marine ecosystems, as well as the existence of European Directives on plastics. Fishermen specifically seem to know about the marine litter problem, but they ignore the impacts on their professional and personal life, especially about the microplastics and how they end up in the food change. Another contradictory outcome of the study was that besides the environmental responsibility was perceived as significant by the majority of tourists, their WTP for environmentally friendly services were comparatively lower to their WTP for a renting better facility. However, when the cost factor was removed from the decision-making choice, the respondents were willing to adapt their needs.

Finally, the project produced a 10-year roadmap identifying the actions that are needed to be taken by the policymakers, the private sector and the society in the future in different time spots to tackle the plastic issue. Top priorities appear to be education and private initiatives in combination with regulation and policy implementation (e.g. circular economy strategy). Education is key to comprehend the damages of plastic and how it affects public health, while the awareness raise could bridge the knowledge gap of adults, who have completed their secondary education. A pioneer in driving change can be the private sector by implementing circular economy models and creating opportunities supported by complementary laws and strategies.

13.2.5 Sustainable Shipping and Ports

Natural challenges related to maritime transport are twofold, to be specific the impacts of maritime transport on the environment (e.g. contamination, CO_2 emanations) and on the other hand the natural effect on maritime transport (e.g. climate change, extreme weather events). Thus, maritime transport and ports play a critical role in addressing the sustainability challenges in international trade, providing market access and linking communities. The European Union has put into force a number of directives and regulations aiming to incentivize port and shipping companies to comply with environmental standards. The new regulatory challenge posed by the sulphur cap in 2020 has generated substantial uncertainty in the shipping industry (IMO 2020). While the shipping industry is focusing on the sulphur cap, the greatest challenge it has likely ever faced is the need to find the effective means of decarbonizing in line with global commitments. The speed of the required transition along with the relative difficulty of technological options vis-a-vis other sectors of the economy makes this a particularly demanding endeavour.

Market-based mechanisms (e.g. emissions cap, emissions trading system) could potentially enact the decarbonizing commitments though they are still far from the centre of the debate. They can motivate the low carbon transition, triggering innovation across CO_2 emission options and providing needed funding both for innovation and supporting developing economies address the burdens of the transition. They are likely however to be one of the many measures, regulations and initiatives needed for the task. Accelerating financial resources and investments will also be an important enabler. This is a role that can be undertaken by regional and national development banks (e.g. the European Investment Bank). Another potential instrument for infrastructural investments is the green bonds.

Chapter 10 provides a brief overview of the marine transport industry, its role and relevance in sustainable development and the kinds of changes that are needed for shipping to be sustainable. It focuses primarily on some of the environmental dimensions of sustainable shipping, although it acknowledges that the effort of sustainable shipping needs to consider several additional aspects to achieve the Sustainable Development Goals. Sustainability is a very broad and sometimes ambiguous concept, but it captures societal values and shapes our vision. Enhancing the sustainability of the maritime transport will require a multi-sector approach involving governments, transport industry, financial institutions, academia and civil society. The inherently international nature of maritime transport seems to make it ideal for global challenges, but it is also a potential weakness given that most governance institutions and their means of enforcing law and regulation are national in nature. A number of government-led initiatives indicate a growing awareness of the shipping challenge, while initiatives at the level of industry and companies suggest a new reckoning of corporate responsibility.

Chapter 11 focuses on the role of ports in implementing CO_2 regulations in international and European waters. The IMO (2020) regulation, bringing the sulphur cap in fuel oil for ships down from 3.50% to 0.50%, is anticipated to bring critical benefits for human well-being and the environment, whereas the European Green Deal, the foremost yearning action plan of the European Union, points at expanding the EU's GHG emissions reduction target for 2030 to at slightest 50% compared with 1990 levels, making the foremost driven bundle of measures, accompanied by a starting guide of key policies in cutting-edge investigate and development, in green innovations and sustainable solutions. Among them, Deep Demonstrations by EIT Climate-KIC using systems innovation approach seek the decarbonization of the European ports and the sustainable transformation of their key components EIT Climate-KIC (2020b). At the same time, international institutions (e.g. IAPH, ESPO) have launched programs pointing at bringing together key partners related to ports to discuss on the most important challenges confronted nowadays, to engage port proactively take the duty of giving reward schemes or green certificates to complied ships and to recognize key markers in measuring GHG outflows.

The study finds out that most of the EU ports are effectively working to secure the environment with the goal of complying with sustainable development standards. Most of the environmental indicators reflect a positive evolution since 2013. In principle, EU ports continue to improve their environmental performance and to maintain or even enhance the declared policies of compliance, environmental protection and sustainable development. Whereas it is troublesome to distinguish and actualize at once 'best practices' for all the environmental impacts that port exercises create, positive steps towards sustainable development and administration are progressively taking place.

13.2.6 Circular Economy

A simple approach to tackle many of the challenges addressed above is integrating circular thinking in the majority (if not all) of the economic activity. The United Nations (UN) Sustainable Development Goals (SDGs) and the EU Circular Economy Action Plan constitute the most prominent efforts promoting the circular economy (CE) influencing/supporting policy agendas in all their members (UN 2020; European Commission 2020c). CE is bringing together the values of sustainability while having an impact on the economy, the environment and the society.

Chapter 12 aims at shed light at the opportunities, which can arise by combing smart specialization (SSS) with circular economy in pilot areas in a European country, which is underperforming in CE implementation, Greece. The methodology used is a policy mapping followed by validation participatory workshops and meeting targeting key stakeholders in the country. The study finds that both CE and SSS need to be revised in order to be more implementation oriented and efficient in financial terms. CE implementation in Greece and other EU countries will be further studied by two more EIT Climate-KIC projects, namely, CL Hub and CE Beacons, which aim at identifying the key drivers of decision-making related to the uncertainty around climate changes and how circular thinking could resolve these doubts, and at implementing and teaching circular approaches to entrepreneurs and investors.

13.3 Directions for Future Research

The Commission develops efforts that create, share and communicate scientific knowledge supporting the aims of the seventh Framework Programme (FP7) Environment and the Horizon 2020 (European Commission 2014, 2020a). Efforts will be undertaken to disseminate and promote the use of relevant knowledge accumulated under recent research and programs, as well as promoting new research in line with priority setting that takes place regularly under the auspices of the Commission. A number of the priority areas identified in the seventh Framework Program for Research and Development and Horizon 2020 could benefit the objectives of Horizon Europe, the next research and innovation framework programme for the period 2021–2027 with € 100 billion budget. Besides theme-oriented international actions carried in H2020, including inter alia projects in the areas of health, food, agriculture and forestry, biotechnology, production technologies, energy, environment (including Climate Change) and transport, Horizon Europe will also target culture, civil security, bioeconomy and natural resources highlighting the importance of innovation in this strategic planning (2020b).

The 2018–2020 Work Program focused efforts on fewer topics with bigger budgets, directly supporting the Commission's political priorities. Four focus areas represent a combined budget of € 7 billion: (i) building a low-carbon, climate resilient future, (ii) connecting economic and environmental gains – the circular economy, (iii) digitizing and transforming European industry and services and (iv) boosting the effectiveness of the Security Union. The focus areas cut across thematic boundaries and bring together contributions from various program parts to pursue a common objective and create sustained impact. They are endowed with a substantial budget to allow for work of sufficient scale, depth and breadth.

From the analysis performed in Chaps. 2, 3, 4 and 5 of this book, it is clear that MUOPs can potentially benefit from each other in terms of infrastructure, maintenance, etc. Knowledge developed over the past years in the projects MERMAID and TROPOS can contribute to a careful integral implementation of MUOPs in Europe, including all relevant stakeholders, in order to overcome obstacles to enhance MUOPs for energy production and aquaculture in the future. The research projects MERMAID and TROPOS have encouraged cooperation among stakeholders and brought together marine sectors such as aquaculture, wind energy, wave energy, mooring and offshore engineering and other blue economy activities to learn and discuss MUOPs. In both the Mediterranean and the Atlantic cases, the efforts made through research projects are the most important, in which the sectors actively have been involved. Through dialogues, increased attention and credibility of MUOPs have developed and is now seen to be relevant to future governance strategies. The research projects also allowed learning among participants, which can be useful to future innovation. Still, the challenges are not overcome, there is a need to be creative and perform sufficient research on this topic. In the future, an improved understanding of how self-governance, network governance and knowledge governance arrangements can be implemented in a strategic and responsible manner will be critical to future MUOP developments.

Chapter 6 presents novel IT applications developed by BlueBRIDGE (H2020) project, which can facilitate producers to engage in the technology race and use it at their own benefit and policymakers to make well-informed and forward-looking decisions. In the future, these platforms, tools and techniques could be replicated and developed by other sectors as well. The preliminary findings yielded in COASTAL (H2020) project, which is presented in Chap. 7 indicate that new, innovative practices such as cooperative activities, alternative forms of tourism, offshore renewable energy and coastal risk management strategies can play a critical role in co-creating a common vision for the development of an interconnected rural-coastal-sea region. The source-to-sea concept bridges the chasm for better integration, cooperation and coordination of activities from the rural area until the ocean aiming at a harmonized and sustainable land-sea area.

Chapter 8 discusses the cultural ecosystem services (CES) frameworks along with the monetary and non-methods for their valuation, supporting that CES valuation can become an extremely useful tool that can bring to the surface the benefits derived from the cultural aspects of MPAs. Integrated and adaptive management will support the MPA administration to adopt best practices, as this opportunity has been undermined so far. Chapter 9 depicts the key challenges of plastic marine litter as identified in the EIT Climate-KIC project, BL.EU. Climate. Tourists and fishermen seem to lack understanding on both the impact of plastic pollution of the seas and European and national regulation. Essential for the understanding of the damages of plastic and the associated effects on public health is the education in all levels of society. The highly needed change can be driven by the private sector through circular economy models and job creation supported by complementary laws and strategies.

From the analysis carried out in Chaps. 10 and 11 it is clear that the maritime transport sector including ports not only are driving up the global temperature (causing 3% of global GHG emissions) and but are an essential part of the global economies (80% of global trade is transported through the oceans). Ports role will be crucial in law enforcement through reward schemes and priority entrance to ships complying with International and European regulation. IMO 2020 regulation in alignment with Agenda 2030 and the SDGs, the European Green Deal and the upcoming European Climate Law are only some of the driving policies towards the decarbonization of the maritime sector. However, they need to put explicit targets and financial plans on the table so that their ambitious objectives will be implemented, as well as revise market-based mechanisms (e.g. EU ETS) including all key drivers of environmental sustainability.

Marine and maritime research for Blue Growth will be implemented through a strategic and coordinated approach across all challenges and priorities of Horizon 2020 and Horizon Europe taking into consideration pivotal concepts, such as the circular economy and the systems innovation approaches. Further research will aim at unlocking the potential of resources from seas, oceans and inland waters for different uses and across the range of marine and maritime industries while protecting the environment and adapting to climate change. Blue Growth will support sustainable growth in the marine and maritime sectors, through sustainable exploitation of marine resources for healthy, productive, safe, secure and resilient seas and oceans. The highly needed sustainability is a well-defined concept, combining economic, environmental and societal values, while it shapes our vision conforming with ethical behaviour that has driven change throughout human history.

References

AMARE. (2020). Interreg MED Programme 2014 - 2020. AMARe – Actions for marine protected areas. Online platform. https://amare.interreg-med.eu

BL.EU. Climate. (2020). EIT Climate-KIC project. Climate Innovation in Southern Waters. https://www.athenarc.gr/el/climate-innovation-southern-european-waters-bleu-climate

BlueBRIDGE. (2020). Horizon 2020 European Commission project. Building Research environments fostering Innovation, Decision making, Governance and Education to support Blue growth. https://www.bluebridge-vres.eu/about-bluebridge. Accessed on 11 February 2020.

COASTAL. (2020). Horizon 2020 European Commission project. Collaborative Land-Sea Integration Platform. https://h2020-coastal.eu

EIT Climate-KIC. (2020a). Europe's leading climate innovation initiative. https://www.climate-kic.org/who-we-are/what-is-climate-kic/. Accessed on 12 April 2020.

EIT Climate-KIC. (2020b). Deep demonstrations. https://www.climate-kic.org/programmes/deep-demonstrations/. Accessed on 12 April 2020.

European Commission. (2012). Blue Growth. https://ec.europa.eu/maritimeaffairs/policy/blue_growth_en

European Commission. (2014). FP7-ENVIRONMENT – Specific Programme "Cooperation": Environment (including Climate Change). https://cordis.europa.eu/programme/id/FP7-ENVIRONMENT

European Commission. (2020a). Horizon 2020 Programme: Funding & tender opportunities. http://ec.europa.eu/research/participants/portal/desktop/en/home.html

European Commission. (2020b). Horizon Europe – The next research and innovation framework programme. https://ec.europa.eu/info/horizon-europe-next-research-and-innovation-framework-programme_en

European Commission. (2020c). EU Circular Economy Action Plan: A new Circular Economy Action Plan for a Cleaner and More Competitive Europe. https://ec.europa.eu/environment/circular-economy/. Accessed at 11 Aug 2019.

IMO. (2020). Sulphur 2020 – Cutting sulphur oxide emissions. http://www.imo.org/en/MediaCentre/HotTopics/Pages/Sulphur-2020.aspx

Koundouri, P. (2017). The ocean of tomorrow, investment assessment of multi-use offshore platforms: Methodology and applications – Volume 1. Springer International Publishing. eBook ISBN: 978-3-319-55772-4, Hardcover ISBN: 978-3-319-55770-0. https://doi.org/10.1007/978-3-319-55772-4.

MERMAID. (2015). Online platform. http://www.vliz.be/projects/mermaidproject/. Accessed 18 April 2018.

MEDFreeSUP. (2020). Single-use plastic free systemic local applications along the Mediterranean east coast, path for a common set of protocols through experiments in Italy, Croatia and Greece. In *EIT Climate-KIC project*. https://www.athenarc.gr/el/medfreesup-single-use-plastic-free-systemic-local-applications-along-mediterranean-east-coast-path. Accessed at 15 Aug 2020.

RECONNECT. (2020). Interreg Balkan-Mediterranean Programme 2014-2020. Regional cooperation for the transnational ecosystem sustainable development. Online platform. https://reconnect.hcmr.gr

TROPOS. (2012). FP7 European Commission project. Online platform. http://www.tropsplatform.eu/

UN. (2020). SDG Indicators. Global indicator framework for the Sustainable Development Goals and targets of the 2030 Agenda for Sustainable Development https://unstats.un.org/sdgs/indicators/indicators-list/. Accessed at 11 Aug 2019.

The manufacturer's authorised representative in the EU is Springer Nature Customer Service Centre GmbH, Europaplatz 3, 69115 Heidelberg, Germany. If you have any concerns regarding our products, please contact ProductSafety@springernature.com

Printed and bound by CPI Group (UK) Ltd, Croydon, CR0 4YY
30/09/2026
02225691-0004